G000254712

Economics and the Challe

Economics and the Challenge of Global Warming is a balanced, rigorous, and comprehensive analysis of the role of economics in confronting global warming, the central environmental issue of the twenty-first century. It avoids a technical exposition to reach a wide audience and is up to date in its theoretical and empirical underpinnings. It is addressed to all who have some knowledge of economic concepts and a serious interest in how economics can (and cannot) help in crafting climate policy. The book is organized around three central questions. First, can cost-benefit analysis guide us in setting warming targets? Second, what strategies and policies are cost-effective? Third, and most difficult, can a global agreement be forged between rich and poor, the global North and South? Although economic concepts are foremost in the analysis, they are placed within an accessible ethical and political matrix. The book serves as a primer for the post-Kyoto era.

Charles S. Pearson is Senior Adjunct Professor of International Economics and Environment at the Diplomatic Academy of Vienna and Professor Emeritus at the School of Advanced International Studies (SAIS), Johns Hopkins University, Washington, DC. During his tenure at SAIS, he directed the International Economics Program for seventeen years and taught at all three campuses in Washington, Bologna, and Nanjing. His teaching and research reflect a deep interest in international environmental economics. He pioneered seminars on trade and environment, the role of multinational corporations, and environmental cost-benefit analysis. His books reflect these interests, with research on global warming published as early as 1978. They include *Environment: North and South*, *International Marine Environment Policy*, and *Economics and the Global Environment* (Cambridge University Press, 2000). He has been Adjunct Senior Associate at World Resources Institute and the East-West Center, and consultant to the U.S. government, international organizations, and industrial, financial, and legal organizations in the private sector. He received his Ph.D. in economics from Cornell University.

Economics and the Challenge of Global Warming

CHARLES S. PEARSON

Diplomatic Academy of Vienna and Emeritus,
Johns Hopkins University

CAMBRIDGE
UNIVERSITY PRESS

CAMBRIDGE UNIVERSITY PRESS
Cambridge, New York, Melbourne, Madrid, Cape Town,
Singapore, São Paulo, Delhi, Tokyo, Mexico City

Cambridge University Press
32 Avenue of the Americas, New York, NY 10013-2473, USA

www.cambridge.org
Information on this title: www.cambridge.org/9781107649071

First published 2011

Printed in the United States of America

A catalog record for this publication is available from the British Library.

Library of Congress Cataloging in Publication Data
Pearson, Charles S.
Economics and the challenge of global warming / Charles S. Pearson.
p. cm.
Includes bibliographical references and index.
ISBN 978-1-107-01151-9 – ISBN 978-1-107-64907-1 (pbk.)
1. Climatic changes – Economic aspects. 2. Global
warming – Economic aspects. I. Title.
QC903.P395 2012
363.738′74–dc22 2011015311

ISBN 978-1-107-01151-9 Hardback
ISBN 978-1-107-64907-1 Paperback

To the grandchildren – Ryan, Emily, Emma, Jack, Grace,
Scott – and to their children, yet to come

The summer is over, the harvest is in, and we are not yet saved.

Jeremiah 8:20

Contents

Acknowledgments   *page* xi

Introduction and a Road Map   1
   Scope and Focus   1
   Motivation and Audience   4
   Structure   5

1. Climate Change: Background Information   9
   The Science   9
   The International Policy Response   14

2. The Role of Benefit Cost in Climate Policy   19
   Background   20
   Inability to Make Secure Inter-Generational Transfers   21
   Willingness *and* Ability to Pay   23
   Risk and Uncertainty   25
   Catastrophe   31
   Sustainability   34
   Alternatives: Tolerable Windows, Safe Minimum Standards,
      Precautionary Approach, and the 2°C Target   37
   Summary   39

3. Discounting and Social Weighting (Aggregating
   over Time and Space)   43
   Introduction   43
   Discounting   44
    Descriptive versus Prescriptive Approaches   45
    The Ramsey Equation   48
    Rho, the Pure Time Preference Parameter   49
    Adjusting for Consumption Growth   52
    Deconstructing Eta   53

	Estimating Eta	57
	Taking Stock	59
	Unsnarling the Discount Rate Tangle	60
	Richer Models	60
	Recalculating Damages	60
	Declining Discount Rates	61
	Social (Equity) Weighting: Aggregating over Space	63
	Concepts	63
	In Practice	66
	Complications	66
	Conclusions	69
4.	Empirical Estimates: A Tasting Menu	73
	Integrated Assessment Models	73
	Damage Functions: The Weakest Link?	76
	An Uncertain Bottom Line	78
	Generating the Numbers	79
	The Art of Shadow Pricing	79
	Agriculture	81
	Sea-Level Rise	83
	Adaptation Costs	86
	Counting (on) Trees: Slowing Deforestation	91
	Conclusion	95
5.	Strategic Responses	97
	The Development Option	97
	Adaptation versus Mitigation	99
	Supply, Demand, and the Green Paradox	106
	Technology	111
	Is Technology Policy Needed?	112
	Mitigation-Technology Connections	114
	Empirical Studies	115
	Geo-Engineering	117
	Conclusions	119
6.	Targets and Tools	124
	Absolute versus Intensity Targets	124
	Certainty of GDP Growth	125
	Uncertain GDP Growth	126
	Choices	128
	The Toolbox	129
	Market Incentives versus Regulation	131
	Emission Taxes versus Cap-and-Trade	133
	International Aspects of Taxes and Cap-and-Trade	137

Subsidies: The Other Market-Incentive Tool   138
Summary   140

7.   Trade and Global Warming   143
    The Impact of Trade on Global Warming   144
      Analytical Approaches   144
      Additional Connections Linking Trade to
        Global Warming   147
    The Impact of Global Warming and Global Warming
      Policy on Trade   149
      Carbon Leakage: Concepts   149
      Carbon Leakage: Estimates and Policy Response   151
      Border Tax Adjustments   153
      Border Adjustments: Legal Aspects   157
    Food Miles, Carbon Labeling, and Other Trade Issues   158
    Carbon Embodied in Trade   160
    Global Warming Policy and the Dutch Disease   164
    Manipulating Permit Markets   167
    Conclusions   168

8.   The Challenge of International Cooperation   171
    Introduction   171
    Concepts: Public Goods, Public Bads   172
    Supplying Global Public Goods   176
      An Ecological Surplus   176
      Self-Enforcing IEAs   179
      Carrots and the Search for Cooperation   182
      Three IEA Simulations   183
      The Dual Role of Transfers   189
      Stepping Back   192
      Sticks   194
    The Need for Cooperation   196
      Cost Considerations   197
      A Shrinking Target Space   200
    Conclusions   202

9.   Beyond Kyoto   206
    From Kyoto to Bali to Copenhagen   206
    From Copenhagen Forward   208
      Cancun   211
      In the Aftermath: The CDM and Sector Agreements   211
      Adaptation Funding   214
    A Limited-Ambition Agenda   215
      Using Price Signals   215

　　　Linking Cap-and-Trade Arrangements　　　　　　216
　　　A Spontaneous Emissions Reduction Credit Market?　218
　　Technology Policy　　　　　　　　　　　　　　219
10.　A Summing-Up　　　　　　　　　　　　　　222
　　Conclusions　　　　　　　　　　　　　　　222
　　Prospects for "Atmospheric Economics"　　　　225
　　Prospects for Climate Policy　　　　　　　　226

Index　　　　　　　　　　　　　　　　　　227

Acknowledgments

I thank Judith Dean and James Riedel for thoughtful and helpful comments. I also gained inspiration and valuable feedback from many students at SAIS (Washington, Bologna, and Nanjing) and at the Diplomatic Academy of Vienna. Thanks to all!

Introduction and a Road Map

An economist's guess is liable to be as good as anybody else's.
Will Rogers, American humorist

Scope and Focus

Global warming is *the* environmental issue of the twenty-first century. Many believe it ranks with war and poverty as one of the greatest challenges to human well-being. But unlike war and poverty, which humanity has confronted for millennia, global warming is a recent concern. And unlike war and poverty, global warming is mainly a prospective threat and one that can in principle be met with pre-emptive action.

Understanding and responding to global warming requires many scientific disciplines including meteorology, climatology, and oceanography; the full array of biological and ecological sciences; and the engineering disciplines. But while science is a necessary component of policy, it is not sufficient.

Global warming presents both old and new political challenges. Measures to limit global warming involve near-term costs with only a promise of benefits, often far in the future. Such actions are inherently difficult for politicians focused on the next election. More fundamentally, virtually all measures to address global warming will affect existing de facto property rights and create winners and losers. And the distribution of the tens of billions of dollars in gains and losses depends on the specifics of policy – abatement targets chosen, economic sectors

1

penalized or subsidized, the market and regulatory tools employed. Politics permeates the rearrangement of property rights.

Confronting global warming is also an international political problem of great complexity and will require statecraft of the highest order. All countries, large and small, North and South, rich and poor, generate greenhouse gas emissions and contribute to the problem, albeit at very different historical, current, and projected levels. At the same time all countries and virtually all groups within countries will be affected by global warming – a few positively, most negatively. The daunting international political challenge is to reconcile these greatly divergent interests and capabilities, and to undertake a potentially costly program of mitigation and adaptation measures, all within an international political system that lacks an international environmental protection agency with the authority to compel emission reductions.

Global warming raises profound ethical issues. The most serious of these is the responsibility of this generation to bequeath to future generations an acceptable environmental inheritance. This question of stewardship is present in many environmental decisions – maintaining wilderness areas, conserving genetic diversity, and the long-term management of nuclear wastes. But the magnitude of our ability – this generation's ability – to affect future well-being through global climate change is unprecedented and raises ethical issues to a new level of concern. What trade-offs exist and what balance should be struck between inter-generational equity and efficiency? What do we owe the future? On the other hand, ethical concerns have a double edge. Should we sacrifice our use of cheap fossil fuel energy today so that generations yet unborn, who presumably will be richer than we are, can avoid adjusting to a warmer world?

Other, more practical ethical questions arise. How should the near-term costs of mitigating global warming be allocated among countries in a fair and efficient fashion? A global effort is needed, but without at least a perception of fairness, governments will not participate. Much the same question arises within countries. Both a concern for social justice and a need to secure political support for mitigation efforts will require some protection or compensation for those that will bear the heaviest abatement and adaptation costs. Ethics are again conflated with efficiency.

The science, politics, and ethics of global warming are not the whole story. This book is primarily about the economics of global warming. Economics offers a powerful set of theoretical and empirical techniques for formulating appropriate responses. But the economics of global warming are not detached from the scientific, political, and ethical dimensions. On the contrary, they are closely linked. Economic modeling of global warming and mitigation policies employs the results of scientific work as a starting point. These combinations of science and economics are known as integrated assessment models and are discussed later. The point here is that economic analysis of the costs and benefits is critically dependent on the underlying scientific research. Moreover, there is a close connection between political analysis and economics in devising global warming policies that are economically efficient and that have some prospect for success. Political economy is central to evaluating the policy instruments and tools to accomplish greenhouse gas abatement. And international political economy is the starting point for analyzing international environmental agreements to limit global warming.

Finally, economics rests on certain value (ethical) assumptions and can help clarify ethical choices. Although economics cannot determine an optimal distribution of wealth and income – an ethical question within the domain of moral philosophy – it can trace out the distributional consequences of policies at a point in time and over future generations. It can also trace the distributional impacts of doing nothing, or following a "business as usual" path. In short, economics can help us understand: Which countries and groups will bear the costs of global warming? Which generations? Are these distributional results equitable? How would various policies change the distributional consequences? The interplay of efficiency and equity comes out most sharply in inter-generational questions. Economics uses the tool of discounting to express future monetary values in terms of present values. It is, in effect, an inter-temporal exchange rate. Discounting has an efficiency objective – the efficient use of resources over time. But as we shall see, it also lies at the heart of the inter-generational distribution of welfare, and hence has an unavoidable ethical dimension.

To summarize, this book is primarily about the role that economics can play in the global warming debate, but it is set within a richer matrix that includes the contributions of science, national and international politics, and equity.

Motivation and Audience

The concept underlying this book is that major events in the world are powerful drivers of advances in economics. The development of national income accounting in the 1930s was closely related to needs created by the Great Depression. Economic planning in World War II contributed to the development of input-output analysis. The burst of public interest in environment in the early 1970s led to major advances in the theory of environmental policy. Events can also overturn conventional economic wisdom. Ricardo wrote of "the inherent indestructibility of the soil," but the Dust Bowl more than 100 years later laid that idea to rest. In the seventeenth century, Grotius, the father of the freedom-of-the-seas doctrine, asserted that the vagrant waters of the sea should necessarily be free as neither navigation nor fishing could exhaust their services. That claim rings hollow with today's fishing technology and fleets.

This book contends that global warming is having a similar impact on economic research. The areas directly affected include discounting and inter-generational efficiency and equity, situating economic systems within an environmental matrix and examining interactions, the design of policy tools in second-best situations, policy formation under extreme uncertainty and potential catastrophe, and our understanding of coalition theory and the supply of global public goods.

These recent advances rest on foundations carefully laid down earlier. We believe that collecting and organizing them in a coherent fashion serves two purposes. First, it underlines how far economics has come and how far it still needs to go to successfully address global warming. Second, much of the recent analysis is appearing in working papers and technical journals or in collected volumes dealing with a narrow slice of the issues and addressed to economist colleagues who are working in this field. It is useful to organize, consolidate, and interpret these advances for those who have not had the opportunity to follow the issues in detail.

We have avoided a technical exposition to reach a wide audience, but have attempted to be accurate and current in terms of presenting the economic underpinnings. Much of the specialized literature relies on mathematical presentation of underlying models and extensive charts and tables to present results. Because this book does not report new research, but synthesizes and interprets recent advances, we have chosen a different route. Our goal is to present complex theory in the simplest fashion possible while respecting the basic logic. We have also summarized the results and policy implications of many different empirical studies and assessed their strengths. For readers who wish to dig deeper, we have included references to the detailed studies on which this manuscript is based. If we are successful, the readers will emerge with an appreciation for the complexities of the economics but also with a firmer foundation for their own beliefs.

Structure

The book contains ten chapters. Chapter 1 starts with a brief review of the science of global warming and of international efforts to moderate climate change. It simply sets a context for readers unfamiliar with the problem and policy initiatives to date. The following chapters are structured around three questions: What amount of global warming is acceptable and what is too warm? What strategies and tools for moderating warming can be deployed? How can we mount a global effort at limiting warming in a world of sovereign states pursuing their narrow self-interest?

Chapter 2 considers whether benefit cost (BC) is an appropriate technique for framing the global warming problem and devising policy. In the BC approach, the benefits of actions to mitigate global warming are the costs averted – the monetary value of future global warming damages that are avoided by reducing greenhouse gas emissions now. The costs of the policy are opportunity costs, the valuable goods and services that the world forgoes by using real resources such as labor, physical and human capital, and technology to reduce emissions. These costs include economic output lost as less polluting but more expensive fuels and energy are used, the costs of sequestering greenhouse gas emissions, and the costs of prematurely scrapping physical capital to reduce emissions. A comprehensive framework also allows consideration of the costs and benefits of *adapting* to global warming, the actions taken to minimize damages occurring when warming takes place. The deceptively simple conclusion from BC – that a policy is justified if the marginal costs of the policy equal marginal benefits, and total benefits exceed total costs, all properly discounted – is shown to conceal many profound complexities. An understanding of the weaknesses as well as the strengths of benefit-cost analysis is needed.

The chapters immediately following elaborate on the benefit-cost approach. Chapter 3 examines the contentious issue of discounting, a procedure that frequently divides economists and environmentalists, but one that also is hotly debated among economists in the context of global warming. As it turns out, the inter-generational equity dimension of discounting is closely linked to the issue of social (equity) weighting – the practice of giving different weights to costs and benefits accruing to individuals at different income levels. Benefit-cost analysis was originally designed to evaluate projects and policies *within* a country and *within* a single generation. But global warming is necessarily *international* and *inter-generational* in scope. This creates additional problems for discounting and social weighting of costs and benefits.

Benefit-cost analysis requires monetary values. In the case of global warming, this means monetary values for the harm (damages) that global warming will produce and for the costs of mitigation or adaptation. Finding monetary values is inherently difficult as many of the effects involve non-marketed goods and services for which there are no market prices to indicate values. Other complications are the high level of scientific uncertainty, the very long time horizons, and our inability to fully anticipate technological advances. In short, it is not surprising that the estimates are contentious. They are, however, central to attempts for a rational policy response to global warming. Chapter 4 explains how the numbers are generated. It is not always reassuring.

Chapter 5 is a transitional chapter. Mitigation – reduction in the emissions of greenhouse gases – is the centerpiece of efforts to control global warming. Putting a price on emissions is at the center of efforts at reduction. However, mitigation takes place within a larger strategic policy space. This chapter considers the broader context, including accelerated development, adaptation, the role of technology, the "green paradox," and the extreme response of geo-engineering.

The two chapters that follow concern policy and institutional arrangements with an eye on economic criteria. Chapter 6 starts by examining the confusing ways in which mitigation targets can be expressed. It then examines the tools available to governments to reduce emissions of greenhouse gases. The principal contenders are the so-called command-and-control or regulatory measures such as vehicle mileage standards to reduce carbon emissions, market-friendly measures such as carbon taxes and cap-and-trade (tradable permits) systems, and various subsidies to accelerate the development of clean technology and renewable energy sources. These approaches can involve very different efficiency and distributional effects that need to be sorted out. Some of these complications involve interactions with existing tax structures, the recycling of revenues in both a tax and in an auctioned cap-and-trade system, the differing effects of uncertainty, and the effectiveness of government mechanisms to induce technological change.

Chapter 7 considers the intersection of climate policy and trade policy. The principal questions center on the international competitive effect of policies to limit global warming, the possibility of "carbon leakage" through international trade, whereby production of carbon-intensive activities shifts to countries with minimal or no abatement program, and the usefulness of trade policy measures to induce or coerce participation in an international mitigation regime. The prospects of carbon leakage and competitive losses, and the general scarcity of tools to forge voluntary international environmental agreements, make trade policy responses attractive but potentially dangerous. Other trade-related issues include measuring the amount of carbon "embodied" in international trade, carbon labeling as a possible trade barrier, international permit trading leading to the "Dutch disease," and manipulations of the permit market itself.

Climate change is global in scope. Chapter 8 approaches it as a complex problem in the provision of a global public good or, alternatively, preventing a public bad. The theory and practice of providing international public goods takes us into considerations of free-riding, extortion, strategic behavior, and game theory. Even though much of the professional literature is abstract and technical, sophisticated modeling using both game theory and integrated assessment models (IAMs) can provide important lessons to inform post-Kyoto negotiations.

The evolution of climate policy through Cancun and its likely direction in the post-Kyoto period is the subject of Chapter 9.

Chapter 10 provides a brief summary, the main conclusions, and prospects.

1

Climate Change

Background Information

This chapter is for readers who are not familiar with the basic facts of climate change and climate change policy. The Fourth Assessment Report (AR4) of the Intergovernmental Panel on Climate Change (IPCC), released in 2007, provides comprehensive information. It consists of a Synthesis Report and reports from three working groups: WG I (The Physical Science Basis), WG II (Impacts, Adaptation, and Vulnerability), and WG III (Mitigation of Climate Change). The fifth Assessment Report is due in 2014.

The Science

The scientific basis of climate change is well established, although many quantitative relations are subject to great uncertainty. Briefly, certain gases emitted into the atmosphere change the earth's energy balance[1] by allowing incoming shortwave solar energy to enter but inhibiting exit of longwave energy. The result is that increases in the concentration of these gases in the atmosphere change the energy balance, resulting in a rise in temperature.

Global surface temperatures are climbing at an increasing rate. Since 1920, the increase has been about 0.78 °C. The linear trend for the past 50 years (1956–2005) of 0.13 °C per decade is nearly twice the rate for the past 100 years. In 2007, the IPCC reported that the eleven of the twelve warmest years on record (since 1850) occurred in the last twelve years (IPCC AR4 2007a).Other evidence includes the annual

---

[1] Measured by radiative forcing (watts per sq meter).

melting rate of glaciers, which has doubled since 2000 as compared to the rates in the previous two decades. The decline in Arctic sea ice has accelerated from 3 percent per decade in 1979–1996 to 11 percent in the past ten years (Füssel 2008).

The principal greenhouse gases are carbon dioxide (CO_2), methane (CH_4), nitrous oxide (N_2O), and a collection of man-made halocarbons. Carbon dioxide accounts for more than 60 percent of atmospheric emissions and is therefore central to any mitigation strategy.[2]

The principal anthropogenic sources of CO_2 emissions are consumption of fossil fuels (about 78 percent of the total) and land use changes, mainly deforestation. About half the carbon released from fossil fuel combustion goes into the atmosphere. Most of the remainder is absorbed by the oceans. There is some evidence that the oceans may be slowing their uptake of CO_2, further increasing the atmospheric burden (Schuster and Watson 2007).

The main sources of methane are solid-waste landfills, coal mining and oil and gas production, wet rice agriculture, and livestock. Sources of nitrous oxides are nitrogen fertilizer, biomass burning, and fossil fuels.

The carbon content of fossil fuels per unit of energy differs. Coal emits about 25 tons of carbon per million BTUs; oil about 20 tons; and natural gas 15 tons. Thus fuel switching is an essential part of mitigation strategy. Unfortunately, coal is by far the most abundant of the world's supply of fossil fuels.[3]

The lifetime of various gases in the atmosphere also differs. It is estimated that 50 percent of carbon emitted today will remain in the atmosphere for 100 years and 20 percent will remain for more than 1,000 years, although there is considerable uncertainty due to the complex carbon cycle.[4] Nitrous oxide has been estimated to have a fifty-year lifetime, and methane's lifetime in the atmosphere is relatively short, at twelve years. Some halocarbons, such as perflurocarbons, will persist for 50,000 years. The global warming potential of the

---

[2] Water vapor in the stratosphere also acts as a greenhouse gas. Variations in its concentration are not well understood.

[3] One ton of carbon is equivalent to 3.67 tons of carbon dioxide.

[4] Archer and Brovkin (2008) state the literature presents ranges from 20 to 60 percent still in atmosphere after 1,000 years. There has been confusion between the residence time of a specific carbon molecule, which may be short due to interchanges among sinks, and how long it will take for the bulge of anthropogenic atmospheric CO_2 to dissipate.

various gases depends on their atmospheric lifetimes and molecular structures and can be made comparable by conversion to a carbon dioxide equivalent measure, CO_2e. The persistence of at least some of the gases in the atmosphere means that what we emit today will have consequences for centuries to come.

Atmospheric concentrations of greenhouse gases have been increasing. CO_2 has increased from its pre-industrial level of 280 ppm[5] to about 390 ppm today, with the most rapid increases in the past fifty years. Methane concentrations today are at 1,774 ppb,[6] more than twice their pre-industrial level. And of course the various halocarbons did not exist before the twentieth century.

These numbers will increase. In business-as-usual (BAU) scenarios (i.e., no effective abatement policy) the U.S. Department of Energy estimates that CO_2 concentrations could reach 700–900 ppm by the end of the century and continue to rise thereafter.[7] Methane would rise from 1,745 ppb in 1998 to 2,000–4,000 ppb in 2100, and nitrous oxides would rise from its 1998 level of 314 ppb to 375–500 ppb by the end of the century.

Historically, the Organisation for Economic Co-operation and Development (OECD) countries contributed 59 percent of cumulative CO_2 emissions between 1900 and 2004, and Eastern Europe, including Russia, contributed another 19 percent. Developing countries made up the balance – about 22 percent[8] (excluding land use and forestry changes). Although not primarily responsible for historical emissions, developing countries, through rapid growth of fossil fuels use, deforestation, wet rice agriculture, livestock, and other activities, are now significant sources of emissions. On a per-capita basis, however, developing countries' contributions are far smaller. For example, per-capita emissions of CO_2 are currently about 20 metric tons in the United States and 2.7 metric tons in China. Nevertheless the United States produces fewer emissions per dollar of Gross Domestic Product (GDP) than does China.

---

[5] ppm – parts per million.
[6] ppb – parts per billion.
[7] These are not projections but plausible scenarios.
[8] Bosetti et al. (2009), citing World Resources Institute data. Both historical and current contributions are traditionally measured on a "production" rather than a "consumption" basis, and thus neglect the role of trade. See Chapter 7.

The proportions for total (not per capita) future emissions will be very different. In 2008, China overtook the United States as the single largest emitter, also topping the collective emissions of the European Union. Developing countries are expected to account for more than 90 percent of the growth of emissions in a BAU scenario over the next twenty years.[9] The implication is that any effective strategy for moderating global warming must include serious efforts at abatement in both large (China, India) and medium-size, rapidly growing developing countries such as Vietnam.

The influential Stern Review (2007) states that if emissions continue and are sustained at today's levels, atmospheric concentrations would be almost double the pre-industrial levels by 2050, and temperatures would eventually rise by 2 °C–5 °C (3.6°F–9°F) on average. For comparison remember that today's global average temperature is about 14 °C. The IPCC estimates that doubling the CO_2e from pre-industrial levels is likely to eventually increase global temperatures by 2 °C to 4.5 °C ("likely" meaning a probability higher than 66 percent), and that values higher that 4.5 °C cannot be excluded. Some have argued that the IPCC estimates were made before the rapid emissions growth in China and India in the first years of the twenty-first century, and therefore underestimate likely temperature increases (Garnaut 2008). One estimate based on Monte Carlo simulations is a median temperature increase of 4.5 °C between 1900 and 2105 under a BAU scenario, with a 99 percent confidence range of 3.0 °C–6.9 °C.[10]

Climate sensitivity – the response of climate to an increase in greenhouse gas concentrations, including feedback mechanisms – remains quite uncertain. Feedback mechanisms that can accelerate warming beyond IPCC estimates include the release of large amounts of methane currently stored in frozen tundra and in the deep-sea bed, as well as the rapid melting of the Arctic ice cap, allowing open-ocean absorption of the incoming solar flux.[11]

---

[9] There is a debate as to whether national emissions should be measured on a production (the current method) or consumption basis. On a consumption basis, which adjusts for the carbon content of trade, China's emissions are sharply lower. See Chapter 7.

[10] von Below and Persson (2008).

[11] On tipping points and thresholds, see Lenton et al. (2008).

The distributional impact of climate change on rich and poor countries will be highly uneven.[12] "Climate change is likely to impact more severely on the poorer people of the world because they are more exposed to weather, because they are closer to the biophysical and experience limits of climate, and because their adaptive capacity is lower."[13] Disproportional impact on poor countries does not mean that rich countries are off the hook. They have greater capital stock and those assets are vulnerable to severe weather events, sea-level rise, flooding, and so on. Disruption of ocean thermohaline currents (e.g., the Gulf Stream) is also a threat.

Temperature increase will have a few positive impacts, mainly felt by temperate-zone countries in the form of lower winter heating costs and longer agricultural growing seasons. In the tropics, where current temperature generally exceeds optimal levels, the effects are almost uniformly negative for economic and social development. One common way to classify these effects is as follows[14]:

1. Sea-Level Rise. The IPCC projects sea-level rise of 18 to 59 cm over the balance of the twenty-first century. However, because of scientific uncertainty, it did not assess the likelihood of these projections or provide an upper bound. Some recent studies are more pessimistic. Pfeffer and colleagues (2008) believe the most likely increase is 0.8 m, but a rise of 2 m this century cannot be ruled out. The melting of the Greenland ice cap could ultimately raise sea levels by 7 m, but not in this century (IPCC 2007b). Loss of the West Antarctic ice could have a similar effect. Sea levels will continue to rise long past stabilization of concentrations and indeed long past stabilization of temperature. The IPCC expects a continuing rise from melting ice and thermal expansion for the next 1,000 years even if CO_2 emissions were to peak and decline over this century.

2. Agriculture. The IPCC projects crop productivity to increase slightly from CO_2 fertilization in middle and high latitudes for temperature increases of $1\,°C$–$3\,°C$, but to decline for greater increases. Productivity is projected to decrease for even small

---

[12] See, for example, Mendelsohn et al. (2006).
[13] Tol et al. (2004).
[14] See IPCC (2007b) *Working Group II, Summary for Policy Makers.*

temperature increases in seasonally dry and tropical, low-latitude regions.

3. Fisheries. The IPCC projects adverse effects for aquaculture and fisheries. This is partly due to essentially irreversible ocean acidification. Subtropical coral reefs are especially vulnerable.

4. Public health. Adverse effects are predicted as a result of malnutrition, deaths and disease from heat waves, storms and flooding, diarrheal disease, and some infectious disease vectors. Temperate areas will have fewer deaths from exposure to cold.

5. Fresh water. Increasing scarcity and seasonality of fresh water supplies, due in part to a decline in snowpack and glaciers, are expected. Drought-affected areas will likely expand.

6. Severe weather events. Hurricanes, local flooding, droughts, heat waves, and the like may become more frequent and severe.

7. Ecosystem disruption. Among other things, loss of coral reefs, loss of biodiversity and genetic resources, and accelerated extinction of species can be expected. In one study, an estimated 15–37 percent of species in sample regions face extinction by 2050 from mid-range climate warming scenarios (Thomas et al. 2004).

The International Policy Response

Although scientific speculation about carbon emissions and global warming date to the nineteenth century, it was not until the 1970s, when scientists were able to measure and confirm increases in atmospheric concentrations,[15] that governments started to respond. The United Nations Environment Programme (UNEP) and the World Meteorological Organization (WMO) created the IPCC in 1989 to provide a scientific basis for policy. In 1992, the UN Framework Convention on Climate Change (UNFCCC) was signed. The objective was to stabilize concentrations at levels that would "prevent dangerous anthropogenic interference with the climate system" (Article 2, UNFCCC). No binding numerical targets were established for

---

[15] Confirmed by observations between the 1950s and the 1970s at Mauna Loa in Hawaii. For the somewhat haphazard "discovery" of global warming as a problem, see Weart (1997).

emissions or concentrations, however. Protection of the climate system was to be on "the basis of equity and in accordance with [parties] common but differentiated responsibilities and respective capabilities" (Article 3, UNFCCC).

The Kyoto Protocol, signed in 1997, was the next major step. Emission reduction targets averaging about 5 percent below 1990 levels were set for Annex I countries, mainly OECD and former Soviet Union nations.[16] Targets are to be achieved in the period between 2008 and 2012. Sixty-five percent of the 1990-level global emissions were included in the original protocol, but this slipped to 32 percent by 2002.[17] No targets were set for developing countries.

Subsequent negotiations were hung up on a number of issues, including credit for carbon sinks and "supplementarity" – the extent to which a country could fulfill its reduction commitments through various flexibility mechanisms. The stalemate was broken in the spring of 2000, when the incoming Bush administration decisively rejected the Kyoto Protocol. The United States gave as reasons for rejection the failure to set targets for developing countries and high costs to the U.S. economy. Ironically, with the United States choosing to sit on the sidelines, it was easier to reach a compromise that favored Canada, Russia, and certain other countries, but which weakened the Kyoto targets (Babiker et al. 2002).

The Protocol came into effect with the ratification by Russia in 2005. It contains three so-called flexibility mechanisms, one of which is of considerable importance to developing countries. This is the Clean Development Mechanism (CDM) by which Annex 1 countries can meet a portion of their reduction commitments by buying project-specific carbon reduction credits from developing countries.

The EU followed up on its Kyoto commitments by creating the European Trading System (ETS), a cap-and-trade arrangement, to take advantage of the flexibility provisions. Evaluations of the ETS are

---

[16] Technically, the designation of Annex 1 countries refers to a group of countries listed in Annex 1 to the UN Framework Convention on Climate Change. Annex B countries are parties to the Kyoto Protocol that took on explicit emissions-reduction obligations. Annex B includes all Annex 1 countries except Turkey and Belarus. We follow common practice and use the term Annex 1 countries.

[17] Nordhaus (2008). The reasons were the rejection of the Protocol by the United States and the rapid increase in emissions by non–Annex 1 countries.

mixed.[18] The great accomplishment is the establishment of a market-based, continent-wide price for carbon emissions. The shortcomings are: (1) for the most part, the emission allowances were distributed for free rather than being auctioned off; (2) allowances were over-allocated, creating windfall profits to certain businesses; (3) only limited sectors within the EU were covered. Free distribution, although generally welcomed by industry, does not generate revenue that could be used to offset other, distortionary taxes – the so-called double dividend. As a result of over-allocation and inept release of data, the price of carbon has been volatile. The EU plans to modify and extend the ETS to include greater use of permit auctions and to include additional sectors, notably transportation.

Discussions of post-Kyoto arrangements were formally initiated at a meeting in Bali in December 2007. That produced a "road map" for negotiations over the two-year period leading up to Copenhagen in 2009, at which time an agreement on a post-Kyoto regime was scheduled to be presented and signed. The Bali meeting did not establish any targets for emission reductions, but it was successful in three areas: (1) developing countries, for the first time, indicated a willingness to consider mitigation[19] plans; (2) there was broad agreement that countries should be able to earn carbon credits by paying for forest protection in developing countries – a step beyond what the Kyoto Protocol allows; (3) a greater interest was shown in adaptation measures, as well as the need to assist developing countries in their disproportionate adaptation burdens.

The meeting in Copenhagen in December 2009 reached a non-binding "Accord" that has no formal standing within the UN system. The Accord calls for quantified emissions targets for Annex 1 countries for 2020 and calls on non-Annex 1 developing countries to take "nationally appropriate mitigation actions." It contains an aspirational goal of holding temperature increase to less than $2\,^\circ\text{C}$. This is thought to be possible by stabilizing atmospheric concentrations of CO_2 e in the range of 450–550 ppm. There is no authoritative analysis showing an increase of $2\,^\circ\text{C}$ is an "optimal" target.

---

[18] For an assessment, see Hepburn (2007).

[19] Mitigation and abatement are used interchangeably in this book.

In early 2010, countries accounting for some 80 percent of greenhouse gas emissions outlined the efforts they were prepared to make to reduce emissions by 2020. China will endeavor to reduce emissions intensity by 40–45 percent by 2020, and India by 20–25 percent (intensity means emissions per unit GDP. With strong economic growth, this implies increasing emissions from these two countries). The EU announced a target level of emissions by 2020 that is 20 percent below its 1990 level (30 percent, contingent on action by others). The U.S. target is in the range of 17 percent below 2005 levels by 2020, but is contingent on passing domestic legislation. Several developing countries submitted targets expressed as reductions from BAU. Who is to calculate BAU levels is yet to be determined.

The Copenhagen meeting also secured financial pledges from developed countries. The "fast track" pledge was $10 billion annually for 2010–2012 and an annual $100 billion by 2020. No commitment was specified for the intervening years, 2013–2019. The funds are to support both mitigation and adaptation, but the allocation was not specified.

The 2010 Cancun Agreements import essential elements of the Copenhagen Accord into the UNFCCC. However, the meeting did not set out a clear path to a binding agreement.

References

Archer, D. and V. Brovkin (2008). The Millennial Atmospheric Lifetime of CO_2. *Climatic Change* 90 (3): 283–97.

Babiker, M., H. Jacoby, J. Reilly, and D. Reiner (2002). The Evolution of a Climate Regime: Kyoto to Marrakesh and Beyond. *Environmental Science and Policy* 5: 195–206.

Bosetti, V., M. Tavoni, C. Carraro, E. DeCian, R. Duval, and E. Massetti (2009). The Incentives to Participate in, and the Stability of, International Climate Coalitions: A Game-theoretic Analysis Using the WITCH Model. *FEEM Nota di Lavoro 64.2009.*

Füssel, H-M. (2008). The Risks of Climate Change: A Synthesis of New Scientific Knowledge Since the Finalization of the IPCC Fourth Assessment Report (AR4). *Background Note to World Bank World Development Report 2010.*

Garnaut, R. (2008). *Garnaut Climate Change Review.* Melbourne: Cambridge University Press.

Hepburn, C. (2007). Carbon Trading: A Review of the Kyoto Mechanisms. *Annual Review of Environment and Resources* 32: 375–93.

Intergovernmental Panel on Climate Change (IPCC) (2007a). *Climate Change 2007: Synthesis Report.* Accessed at http://www.ipcc.ch/publications_and_data/ar4/syn/en/mains.html

   (2007b). *WorkingGroup 2: Impacts, Adaptation and Vulnerability.* Accessed at Http://www.ipcc.ch/publications_and_dataar4/wg2/en/contents.html

Lenton, T., H. Held, E. Kriegler, J. Hall, W. Lucht, S. Rahmsdorf, and H.J. Schellhuber (2008). Tipping Elements in the Earth's Climate System. *Proceedings of the National Academy of Sciences of the US* 105 (6): 1786–93.

Mendelsohn, R., A. Dinar, and L. Williams (2006). The Distributional Impact of Climate Change on Rich and Poor Countries. *Environment and Development Economics* 11 (2): 159–78.

Nordhaus, W. (2008). *A Question of Balance: Weighing the Options on Global Warming Policy.* New Haven, CT: Yale University Press. Prepublication version at http://nordhaus.econ.yale.edu/Balance_prepub.pdf

Pfeffer, W., J. Harper, and S. O'Neel (2008). Kinematic Constraints on Glacier Contributions to 21st Century Sea Level Rise. *Science* 321 (5894): 1340–43.

Schuster, U. and A.J. Watson (2007). A Variable and Decreasing Sink for Atmospheric CO_2 in the North Atlantic. *Journal of Geophysical Research* 112–22.

Stern, N. (2007). *The Economics of Climate Change: The Stern Review.* Cambridge: Cambridge University Press.

Thomas, C. et al. (2004). Estimating Risk from Climate Change. *Nature* 427 (6970): 145–48.

Tol, R.S.J., T. Downing, O. Kuik, and J. Smith (2004). Distributional Aspects of Climate Change Impacts. *Global Environmental Change, Part A* 14 (3): 259–72.

von Below, D. and T. Persson (2008). Uncertainty, Climate Change, and the Global Economy. *NBER Working Paper* 14426.

Weart, S. (1997). The Discovery of the Risk of Global Warming. *Physics Today* 50: 34–50.

2

The Role of Benefit Cost in Climate Policy

Benefit-cost (BC) analysis comes in two flavors. The standard, "vanilla" flavor examines the monetized costs and benefits of a project or policy. If the benefits exceed the costs, the project is cleared to proceed. This approach, however, is best suited for projects for which there is only one feasible scale or level of intensity. More frequently, there is a choice of scale, and the objective is to maximize net benefits. This more exotic flavor implies two analytical steps – to calculate the scale or intensity where marginal (incremental) benefits equal marginal (incremental) costs, and then to check that at this scale, benefits exceed costs.

Benefit-cost analysis of global warming policy is done using both approaches. Some studies select a target in terms of greenhouse gas emission levels, atmospheric concentrations, or temperature change, and calculate the costs and benefits of attaining the target.[1] In contrast, some studies attempt to calculate the level of emissions such that marginal abatement costs equal marginal benefits, and social welfare is maximized. This latter approach is more difficult as it requires knowledge of costs and benefits over a range of abatement levels. Whichever approach is taken, the least-cost, or most cost-effective, available abatement measures should be examined and selected. And whichever approach is used, costs and benefits should be converted to the same time period, implying discounting.

This chapter and the next consider whether BC analysis is a useful approach to forming global warming policy. We conclude that it is, but should be used carefully and with full awareness of its weaknesses. For

[1] Costs are mitigation (abatement) costs; benefits are damages avoided.

expositional convenience this chapter considers several characteristics of climate change that challenge conventional BC analysis, and defers to the next chapter the challenges that arise from discounting and social (equity) weighting.

Background

Neither the 1992 United Nations Framework Convention on Climate Change (UNFCCC) nor the 1997 Kyoto Protocol rests on a BC foundation. The UNFCCC objective is to stabilize greenhouse gas concentrations in the atmosphere "at a level that would prevent dangerous anthropogenic interference with the climate system." Costs and benefits were not explicit in setting this objective, although members were admonished to take policies that are cost effective (i.e., least cost). The Kyoto Protocol established mandatory greenhouse gas emission targets for Annex 1 countries for the first commitment period (2008–2012). But there was no attempt to justify the aggregate Annex 1 reduction target, averaging 5 percent below 1990 emission levels, or the allocation of emission reductions among countries, on a BC basis. However, as described in Chapter 1, the Protocol did provide for three flexibility mechanism – emissions trading, joint implementation, and the so-called Clean Development Mechanism (CDM) – all of which contribute to the least-cost objective.

The lack of a BC justification may appear odd. At first glance, BC has many attractive features and would appear to be an almost ideal guide to setting policy. It has come to be one of the most powerful and widely used economic techniques to evaluate public expenditure and public policy. It is grounded in mainstream economic theory (welfare economics); it champions efficiency; it can determine the optimal scale and timing of a project; through shadow pricing it can accommodate distorted market prices, and find implicit prices for environmental goods and services that do not pass through markets. BC analysis can be extended to trace out the distributional or equity consequences of projects and policies. It has techniques to deal with risk and uncertainty. Good BC analysis will be explicit about assumptions, and can provide a single number, the net present value (NPV), to decision makers, or, if preferred, a range of values together with their probabilities. Finally BC can provide an overarching framework for linking climate, environmental energy, and economic modules – exactly what is needed in Integrated Assessment Models investigating climate change.

On second glance, however, there are particular features of the global warming *problématique* that tend to call into question the appropriateness of BC analysis. Briefly, these features are the exceptionally long time horizon, the high degree of scientific and economic uncertainty (not unrelated to the centuries-long timescale), the scale of the damages that may arise, and the global character of the challenge and response. We will argue that it is not necessary to completely reject BC as a tool in formulating climate policy, but it is important to understand its limits. We analyze three problem areas, the first arising mainly from the inter-generational time horizon, the second arising from the international character of global warming, and the third arising from pervasive uncertainty. We then consider alternatives to BC. The discussion of discounting is deferred until Chapter 3, but we note here that it too is can be considered a weakness in BC analysis of climate change. It is noteworthy that many of the controversial aspects of BC analysis discussed in this and the next chapter involve fairness or equity within and between generations.

Inability to Make Secure Inter-Generational Transfers

The first serious challenge to using BC for global warming is an esoteric and often overlooked compromise called the Kaldor–Hicks (KH) hypothetical compensation test. Here is the problem. Late in the nineteenth century, an Italian economist, Vilfredo Pareto, laid the primary foundation stone for modern welfare economics by declaring that a state of affairs was *optimum* (efficient) if resources were allocated such that it was not possible to make one person better off without harming another. Note that many Pareto optima are possible, indeed one for every distribution of income. Note also that the word optimum does not imply equitable or fair. It follows that a Pareto *improvement* is a movement toward a Pareto optimum in which at least one person is made better off, and no one worse off. The second foundation stone of modern welfare economics was the recognition that economists, acting as economists, cannot make interpersonal welfare comparisons. The implication of these two propositions is that a government-sponsored policy or project unambiguously improves social welfare only if at least one person is made better off and no one is harmed. Alternatively stated, economics cannot assert an improvement in social welfare even if 100 are made better off and only one is harmed.

This poses an almost impossible hurdle for government action. Virtually all policies create winners and losers. As a practical matter, losers cannot be fully compensated (left unharmed). At this point it appeared that BC as a tool for evaluating public policy was paralyzed. The resolution of this dilemma, which was suggested many decades ago and which undergirds BC to this day, is the KH hypothetical compensation test: Does the policy or project in question generate sufficient benefits so that the winners *could* compensate the losers and still have something left? If so, the policy is justified *even if compensation is not paid*. The practical effects of the KH test are huge. Economists could concentrate on their principal concern, efficiency, and could finesse equity (distributional) questions. Projects and policies could be accepted or rejected on their net benefits, without regard for who received the benefits and who bore the costs.

The justification for this sleight of hand was the presumed existence of a political system and institutions that could separately attend to society's distributional aspirations. It was aided by a well-known welfare theorem that states that in a competitive equilibrium, redistribution to achieve society's equity objectives does not need to conflict with efficiency. Thus efficiency could be hived off from equity. On a more practical level, it was also argued that individuals are affected by many government policies and projects, sometimes positively, sometimes not. If care were taken that all policies passed the KH test, it is likely that for any one individual, the sum of the benefits he or she received from the many projects was larger than the harm – this individual is compensated for damages incurred from one policy by benefits received from many others.

Not all economists are persuaded that the compromise engineered in the 1930s stands up in the environmental era that began in the 1970s. Farrow (1998) bravely argues that in the current *zeitgeist*, concern for environmental equity and sustainability requires BC to meet two additional tests: that actual compensation be paid to groups damaged by pollution or the extraction of natural resources; and that for sustainability, resource rents be reinvested. Although his argument is developed in a domestic context, it resonates strongly in an international context.

Whereas the justifications for KH may be persuasive with short- and medium-term projects within a democratic society, the intergenerational and global character of climate change poses a direct challenge to the test and hence to the use of BC. Simply put, there is no robust inter-generational political system to attend to distributional objectives. The KH justification separating efficiency and equity is eroded. Specifically, there is no political institution or mechanism through which the present generation can securely compensate generations in the far future for the consequences of global warming. All attempts to set up a sinking fund for compensating future victims are subject to plunder in intervening decades.[2] The inter-generational welfare transfer problem is symmetrical. There is no obvious way for future generations, most of whom are likely to be wealthier than we are, to compensate us for our sacrifices if we take expensive greenhouse abatement measures today.[3]

Thus the potential for compensating losers, which is at the heart of the KH test and conventional BC analysis, and which allow separation of efficiency from distribution of gains and losses, is compromised. In Mishan's terms, we confront a *potential* potential Pareto improvement (Mishan 1988). The net effect is not enough to scrap BC but to acknowledge that inter-generational equity concerns must be explicitly addressed together with inter-temporal efficiency in forming global warming policy. As shown in Chapter 3, this directly affects the social rate of time preference.

Willingness *and* Ability to Pay

The importance of explicitly considering equity in global warming analysis is confirmed from another direction. BC analysis needs monetary values. The most fundamental indication of the value we place on goods and services is our willingness to pay (WTP) for them. Many

---

[2] Investment in technology is probably least vulnerable.
[3] Future generations can compensate the current generation for private goods with transferrable ownership rights, but not for public goods such as the climate.

times, our WTP is directly measured by market prices. Indeed, prices themselves reveal aggregate WTP. In some instances, and especially for non-marketed environmental goods and services, WTP must be inferred from consumers' behavior in surrogate markets, or from direct surveys (see Chapter 4). In other instances, when confronted with the prospect of losing a good or service, value is measured by willingness to accept (WTA) compensation for the loss of that good.

WTP and WTA are deceptive terms, however. Both are constrained or affected by ability to pay, or income. In fact, the distribution of income among individuals will determine the aggregate WTP and hence the explicit or implicit prices of the goods and services. As prices are used as a measure to monetize benefits and costs, it follows that different income distributions will generate different prices, different monetary values for benefits and for costs, and hence different BC ratios. Multiple BC ratios undermine BC as a decision tool. Which ratio is dispositive? In particular, if the existing distribution of income is widely considered unjust, the results of BC analysis based on that distribution are tainted. The resolution to this dilemma is not unlike the compromise underlying the KH hypothetical compensation test. It is asserted that in a democratic society with recourse to functioning political institutions and redistributive tax systems, one can assume that the existing distribution of income is "just" or "fair." Thus the set of prices on which BC is based can be considered fair, or at least untainted.

Once again, recourse to a political system to ensure a fair distribution of income stumbles when it comes to global warming. There is no world government through which democratically determined equity objectives are attended to. Very few would argue that current world income distribution is just or fair or equitable. But it is the existing and prospective income distributions, and the consequent structure of world prices, that are inputs to BC models. And it is the output of these models that support either aggressive or minimal response to impending climate change. In short, if left unadjusted, BC analysis of global warming reflects world income distribution patterns that are widely considered unfair.

The problem can be illustrated in a very specific context. It is widely agreed that global warming will cause additional deaths and these deaths will occur disproportionately in poor countries. The standard way to monetize increased mortality is to estimate the value of a statistical life (VSL). Estimates of VSL in the United States, derived mainly from hedonic wage models, tend to cluster around $6 million (Viscusi and Aldy 2004).[4] If this value is scaled by the per-capita income difference between India and the United States, an estimate of WTP for saving a statistical life in India would be less than $500,000. If this value were then used to project the monetary cost of Indian deaths, we would be left in the ethically uncomfortable position of low-balling global warming damages, weakening abatement policy, and increasing deaths in poor countries, all because damages in poor countries are given a low monetary weight. Indeed, the larger the fraction of damages borne by poor countries, the lower the monetized damages, the weaker the abatement policy, and the more damages they will suffer. The root problem is, of course, the absence of a functioning international political system that can address inequalities in income. All this suggests that some deliberate weighting of the incidence of global-warming damages – who bears the costs – may be necessary. Social weighting can address this problem and is considered in detail in the next chapter. It turns out that the case for weighting is more complex than it appears on the surface. Here it is sufficient to note that the equity dimension of climate policy again intersects in an essential fashion with efficiency considerations, and BC analysis must be sensitive to both.[5]

Risk and Uncertainty

Risks and uncertainties pervade the science and economics of climate change. In itself this is not news.[6] Regardless of sector, BC is always

---

[4] Hedonic wage models attempt to estimate the value of a statistical life based on wage differentials for occupations with high and low risk of fatality.

[5] To anticipate, it may be argued that there are perhaps more efficient ways to improve the plight of the poor than manipulating VSL. The counterarguments elaborated in this and the next chapters are the inadequacy of international and inter-generational mechanisms for compensating global-warming victims (i.e., the weakness of the KH compensation test), the disproportionate role of the rich in creating the problem, and consistency with inter-temporal social weighting in conventional discounting practices. These considerations make global-warming damages a special case.

[6] Economics often makes a distinction between risk, where probabilities of outcomes are known, and uncertainty, where probabilities are not known and perhaps unknowable. It is helpful to think of risk and uncertainty as end points along a spectrum.

forward looking and hence always confronts risk and uncertainty. The issue at hand is whether the uncertainties are so pervasive and profound as to render BC unreliable and indeed likely to lead us into major policy errors. The special features that distinguish uncertainty in global warming are the presence of non-linearities, thresholds and potential tipping points, irreversibilities, and the long time horizon. The last feature makes projections of technology, economic structure, preferences and a host of other variables 100 years from now increasingly questionable.

Where do the uncertainties cluster? The effects of climate change are frequently set out in quantitative and physical fashion – meters of sea level rise, increased incidence of tropical disease, crop failure due to heat stress, and so on. To appreciate the full scope of uncertainties surrounding these estimates, one has to trace them backward, to the scientific/economic modules in the Integrated Assessment Models (IAMs) from which they were derived, and then trace them forward to their effect on production and utility functions, and from that to market and shadow prices, and to monetary values. A simple listing of the main analytical steps in this chain underscores the multiple sources of uncertainty. Keep in mind that many of these estimates are for the far future, perhaps a century or more from now.

1. Baseline greenhouse gas emissions projections, derived from estimates of population increase, economic growth, and energy composition and use, are needed. For carbon, the Kaya Identity can be employed, where carbon emissions are the product of population times per capita GDP times energy intensity of GDP times the carbon efficiency of energy. Global emissions should be built up from regional or national estimates. Studies have shown that a significant share of the uncertainty surrounding temperature increase is due to socioeconomic drivers: population, total factor productivity, energy efficiency, and land use changes (von Bulow and Persson 2008).

2. The relation between emissions of greenhouse gases (a flow) and atmospheric concentrations (a stock) must be specified. This includes estimating the role of alternative carbon sinks such as the oceans, mixing rates among sinks, and the atmospheric decay rates for various greenhouse gases.

3. The radiative forcing potential of various gases has to be established and transformed into a carbon equivalent metric.

4. The effects of ambient atmospheric concentrations of greenhouse gases on climate (average, seasonal, and regional temperature, rainfall, storms, etc.) are needed. This is the climate system response to sustained radiative forcing, sometimes termed climate sensitivity, and is a major source of uncertainty.

5. The effects of climate change on variables of economic interest are needed. In principle, these should be regional or perhaps local, not global, and should be time-dated. Obvious candidates are sea level change, fresh water availability, crop and forestry yields, fisheries, built structures, health, and tourism. Less tractable but perhaps of equal or greater importance are amenity damages, loss of option values due to irreversibility, species extinction and loss of genetic diversity, and possible interference with beneficial ecological functions such as nutrient recycling and hydrological systems. To the extent possible, all of these effects need to be monetized.

6. Some of the negative effects can be damped or eliminated through adaptation measures, which need to be identified (e.g., seawalls, altered cropping patterns, public health measures). In principle, the cost of these measures should be monetized so that the full damage estimates include adaptation costs and residual damages.

7. Global warming may have positive monetary value, for example improved productivity for temperate zone agriculture or lower winter heating cost, and these should be counted as benefits (negative damages).

At this point, if a BC analysis of a specific mitigation target is at issue, for example either a time path for emission reductions or a ceiling on atmospheric concentrations, the mitigation options have to be identified and costed. Assumptions with regard to how efficiently the mitigation is to be accomplished and how the mitigation burden is allocated among sectors and countries are needed. Some mitigation measures may reduce collateral damages, for example from local air pollutants such as particulates and tropospheric ozone. The monetary value of these avoided damages should be entered as a reduction in mitigation costs, or an addition to mitigation benefits.

If instead of specifying a target, an "optimal" dynamically efficient policy is desired, the next step would be to use mathematical techniques to calculate and project forward the time path of emissions reductions from a business-as-usual (BAU) path that would maximize discounted utility.[7] In either case, it is clear that we would wish to know something about future technologies for mitigation and adaptation and their costs, the future structure of the world economy and its regional variations, and future preferences. These are formidable tasks, and attempts to find defensible numbers are addressed in Chapter 4. The compounding of uncertainties at every link in this chain is a major challenge to BC analysis.[8]

A number of techniques have been developed to deal with risk and uncertainty in BC. Will they be adequate for climate change? The most common is sensitivity analysis, in which an uncertain value for a variable is changed by modest but plausible amounts, and the results, generally expressed as NPV, are examined. If the sign of NPV does not change, there is little point in refining that estimate with better data. Sensitivity analysis helps weed out unimportant assumptions and allows the analyst to concentrate on the key ones. The weaknesses are in judging what is "plausible" and in neglecting any knowledge of the probabilities of outcomes. The expected NPV criterion is a bit more sophisticated. Expected value is calculated by weighting all project (policy) outcomes (measured by NPV) by their probability. Probabilities can be derived from objective data (i.e., rainfall records) or subjectively from expert opinion. If two variables are correlated, their covariance is needed. In general, this criterion requires the expected NPV to be positive for the investment or policy to proceed.

Monte Carlo, a simulation technique, takes uncertainty analysis to the next level. The analyst specifies probability distributions of the uncertain components of the policy or project and specifies the correlations between the components. A computer program then makes

---

[7] Recall that in economics, "utility" connotes the capacity to satisfy wants, and not its everyday meaning of "usefulness."

[8] It remains unclear exactly how sensitive policy advice is to the cascading uncertainties. In a simplified cost-benefit framework, Ralb et al. (2009) found the optimal emission target is quite insensitive to the parameters underlying the cost and benefit (damage) functions. This suggests that despite the pervasive uncertainties, cost-benefit analysis can still be a useful guide to policy.

successive draws on the data subject to the probability distributions, and with a sufficient number of draws builds up a probability distribution of the NPV. The distribution will show the expected value and variance of the simulated distribution, including the risk that the project will have a negative NPV. Even though Monte Carlo analysis provides a better picture of risk, the quality of the results depends critically on entering accurate probability distributions and specifying correct relations between the variables. Although Monte Carlo analysis is useful in mapping out risk profiles, it does not yield a new decision criterion that goes beyond expected NPV.

Neither expected NPV by itself nor expected NPV augmented with Monte Carlo analysis take into consideration risk preferences. The presumption in BC analysis is that for most projects, the government can be risk-neutral – it should neither seek nor avoid risk. This relies on the assumption that most projects are "small" relative to the economy, and that government can spread risk over the population, whereas an individual might well be risk-averse and demand a premium for undertaking a risky project.[9] The presumption of risk neutrality breaks down when the project is "large" relative to the economy, where the impacts are highly concentrated by region or income group and redistribution opportunities are limited, and for projects that provide public goods. In these situations, the literature suggests calculating a risk premium to adjust the expected NPV, with that premium determined in part by an arbitrary relative risk aversion coefficient.[10] Climate change is certainly "large," will affect specific regions and income groups disproportionately, and involves a public good. BC studies of global warming that simply use the expected NPV as the appropriate criterion for action without considering risk preferences are questionable.

An alternative to adding a risk premium to the expected NPV is to use expected utility theory. This is the utility attached to an outcome multiplied by the probability of its occurrence. Utility functions in principle can capture society's aversion to risk. In this way the risk and the utility attached to monetary outcomes can be accounted

---

[9] The Arrow-Lind Theorem.

[10] See Belli et al. (2001). The risk aversion coefficient also enters into the discount rate issue and is considered in Chapter 3.

for, although information on probabilities is still needed. Risk aversion is closely tied to the discount rate, and is considered in the next chapter.

The implication of uncertainty, thresholds, tipping points, and irreversibility is that we should take a precautionary approach. If there is a possibility that better information will become available, it is desirable to avoid taking steps today that lead to irreversible changes, as that information may show the steps to be unwise and expensive.[11] This is the positive value associated with maintaining options.[12] BC analysis can accommodate option values but they are difficult to document and often neglected. Moreover, for option values to actually materialize they require learning, and that implies research. The conventional wisdom is that inclusion of option values supports early and substantial mitigation. A moment's reflection reveals, however, that a precautionary approach does not necessarily suggest greater emissions reductions. It is true that global warming will likely cause irreversible damages in the form of species loss, rising sea levels, loss of agricultural land, and so forth. These are costs that cannot be recovered once they occur and will be the "sunk costs" of global warming. However, efforts to reduce emissions such as scrapping energy-inefficient capital stock and switching to renewable energy sources also involve sunk costs. If in the future, uncertainty is resolved and global warming is less serious than anticipated, these resources have been misallocated. There are sunk costs to investment in abatement as well. In short, irreversibilities occur on both the cost and benefit side, and this muddies the notion of a precautionary policy.

There is another potential irreversibility that is of concern, and that is atmospheric concentrations of greenhouse gases. Some gases have short residence time in the atmosphere, decay quickly, and their concentration is reversible. Carbon, however, has a low atmospheric decay rate and its concentration is only reversible over a long time period

---

[11] Notice that irreversibility only matters for policy if there is uncertainty (Pindyck 2007). If we knew for certain what value the future will place on, say, polar bears, we could more easily put the loss of that species in our cost column and calculate the BC ratio of preventing that loss.

[12] The seminal work was Arrow and Fisher (1974). The precautionary approach is sometimes interpreted as avoiding any action until the full consequences are known with certainty. This is tantamount to using infinite option values and, of course, paralyzes policy.

(Chapter 1). The effects of climate change are uncertain, and we may at some point in the future wish to have lower concentrations than actually exist at that time. As a practical matter, however, emissions cannot be brought below zero in the short term at reasonable cost, at least with current technologies.[13] The (then) existing concentration becomes a quasi-irreversible constraint. This is exacerbated by inertia in the earth's systems, especially absorption of heat by the oceans. This inertia means that the temperature experienced at a point in time is the result of greenhouse gas concentrations that existed in the past. If is, of course, impossible to go back and reduce past emissions. The most we could do is attempt to reduce current concentrations and thereby hold *future* temperatures below where they otherwise would be. These considerations suggest modifying the BC calculus toward greater abatement early on.[14]

Even if the atmospheric concentration constraint does not "bite" in the future, early abatement investments may be justified to minimize costs, especially for long-lived capital, but this must be set against potential cost reductions from (uncertain) advances in abatement technology that can only be captured by research and waiting.

Catastrophe

BC is not well suited for making catastrophe policy. Catastrophes have two characteristics: extreme uncertainty and extreme consequences. BC has difficulty with monetizing extreme consequences, which, because they are large, fall outside the conventional partial equilibrium framework, and because they are rare. Still, extreme consequences are not as serious an analytical challenge as is extreme uncertainty, which is the absence of objective or subjective probability distributions for critical parameters, and especially those relating greenhouse gas concentrations to temperature, temperature to other attributes of climate, and those attributes to damages.

---

[13] Massive aforestation efforts would work in the right direction but perhaps too slowly to accomplish negative emissions. Coupling carbon capture and storage technology with the combustion of biomass fuels (not fossil fuels) also has the potential for bringing atmospheric concentrations below current levels.

[14] For elaboration of the shrinking window for stringent concentration targets see Chapter 8.

Some candidates for triggering climate catastrophe were listed in Chapter 1 and include alteration of the thermohaline ocean circulation systems such as the Gulf Stream, which warms Western Europe, and large-scale melting of the Greenland and West Antarctic ice sheets. In addition, feedback channels that amplify warming are relevant. These include melting of the Arctic ice cap, possible release of methane currently sequestered in permafrost soils, the release of methal hydrates from deep ocean waters, and, with higher temperatures, reduced capacity of the oceans to act as carbon sinks. By and large, Integrated Assessment Models have given little or no attention to feedback mechanisms.

Two studies demonstrate the challenge that catastrophe poses for conventional BC. The first (Tol 2003) starts from the observation that to conduct (expected) BC analysis, the variance of both marginal costs and marginal benefits must be finite. If the variance is infinite, the expected net benefit cannot be calculated and BC loses legitimacy. How is infinite variance of discounted marginal global warming damages possible? The intuitive argument goes as follows. The effects of global warming are uncertain. There is some small but positive probability that for some region of the world growth could become negative. With declining income and consumption, the present value of future consumption is *larger* than its future value, implying a negative discount rate. Moreover if income is equity weighted[15] the fate of the increasingly poor region is given additional weight in the expected present value calculation of global warming damages. In this fashion, the unlikely scenario comes to dominate the expected present value of marginal damage costs and the variance of damage estimates becomes infinite. Given the model, conventional BC is not possible.[16] Underlying this is the model's assumption, widely found in the literature, that utility is equal to the logarithm of per-capita consumption. As per-capita consumption approaches zero, marginal utility approaches infinity.

---

[15] Chapter 3 discusses social or equity weighting.

[16] This may be unduly pessimistic. In a subsequent contribution, Tol and Yohe (2007) add a second policy tool, foreign aid to the afflicted region, which helps rescue income and ultimately rescues benefit cost. In the particular run of the Monte Carlo simulation, it was inadequate water supplies that led to the collapse of income in Central and Eastern Europe. Presumably this could have been foreseen and preventative measures taken.

The second study, dubbed the "dismal theorem" (Weitzman 2009a), poses a more fundamental test for BC. It has come to be known as the "fat tails" problem because the right-hand side of the probability density function may not narrow rapidly. The theorem considers low-probability, high-impact consequences of global warming. The crux of the argument is that extreme uncertainty about critical parameters such as climate sensitivity (the response of climate to temperature increase), combined with high and uncertain damage functions, can result in the probability of a global warming catastrophe falling more slowly than the impacts rise. The expected value of the impact (probability times impact) is then unbounded, and we confront a catastrophic global loss of income and welfare. Ultimately the theorem rests on the idea that we are unable to learn enough about the likelihood of catastrophe from observation until it is too late.[17] Our knowledge of the probabilities of extreme change is inadequate, and we cannot learn enough about them from inductive experience alone to describe them. "The underlying sampling-theory principle is that the rarer is an event the more unsure is our estimate of its probability" (Weitzman 2007, p. 4). We have too few examples of climate catastrophe to predict them.

The implications of this theorem for traditional BC analysis, even BC based on expected utility theory, are profound. The uncertainty problem dominates the analysis so that the issue of discounting, long considered the crux of climate policy analysis, becomes of second-order importance. Weitzman concludes that profound uncertainty places severe limits on policy advice based on standard cost-benefit analysis. It also provides additional support for the precautionary approach.[18]

---

[17] This is known as a Bayesian learning approach. We cannot know the variance of uncertain growth due to global warming from examining historical data. It should be noted that Weitzman assumes a utility function that goes to negative infinity when approaching "catastrophe." A different utility function need not have this characteristic.

[18] Weitzman (2009b) also concludes that the possibility of catastrophe supports research on fast acting geo-engineering responses that could offset and bring down runaway global temperature – an idea discussed in Chapter 5. Dietz (2010) has tended to confirm the dismal theorem using a probabilistic Integrated Assessment Model with "fat tails" in probability distributions, but provides a partial rescue of BC features such as discounting by placing an upper bound on catastrophe damages. Pindyck (2010) also argues that the fat-tail problem recedes if an upper bound is placed on the marginal utility of consumption as consumption levels approach zero.

It does not, however, answer the question of how much caution is warranted.[19]

Sustainability

Studies focusing on uncertainty fall within a larger literature critical of BC on the grounds that it neglects sustainability criteria. Bluntly stated, the charge is that traditional BC places no value on sustainability.[20] How does this relate to global warming? The literature on sustainability is perhaps even larger and more confused than global warming, but there is widespread agreement that a central distinction is between weak and strong sustainability. A weak sustainability rule requires that the total per-capita stock of capital of all types – physical, human, natural (environmental), and perhaps social – are maintained so that incomes can be maintained. However, weak sustainability assumes there is a high degree of substitutability of physical and human capital for natural capital (environmental services) in both production functions and utility functions. Drawing down natural capital through global warming damages or by other means can be offset by accumulation of other forms of capital, and real per-capita incomes can be sustained. In contrast, a strong sustainability rule rejects the easy substitutability between natural and other forms of capital. In conventional economic terms, natural and produced capital are complements, not substitutes, and sustainability requires maintaining at least some minimum level of "critical" natural capital.

---

[19] Pindyck (2009), using information from studies reported by the IPCC and from various IAMs, estimates the willingness to pay a permanent tax on consumption to decrease the likelihood or size of a climate catastrophe. His model suggests that WTP decreases as the pure rate of time preference increases or an index or risk aversion increases. Both work to reduce the present value of future catastrophic damages and hence the willingness to pay for avoiding these damages. For "conservative" parameter values, the willingness to pay to avoid catastrophe is less than 2% of consumption. This analysis does not address the heart of Weitzman's concern – that we do not and perhaps cannot know the probability distributions of critical variables.

[20] Note the irony. Almost forty years ago, the limited *availability* of carbon-rich fossil fuels was perceived as a major threat to sustainability in the influential, if much criticized, study, *The Limits to Growth* (Meadows et al. 1972). In the nineteenth century, the eminent economist, Stanley Jevons, saw scarcity and cost of coal bringing economic stagnation to Britain. Today, the threat is the limited capacity of the atmosphere to *assimilate* the carbon and render it harmless.

The issue of sustainability is obscured in most BC analyses of global warming because they rely on one sector models without explicit consideration of the contribution of environmental capital in either production or utility functions. The implicit assumption is that a loss of environmental capital via global warming can be easily substituted for by produced capital. A few studies, however, have modeled a two-sector economic structure, and the BC results they produced can be quite enlightening. Sterner and Persson (2008), for example, suggest that the services of natural capital – environmental services – may become increasingly scarce relative to conventional goods and their relative price will increase over time. Indeed, the environmental share of the economy can grow in value terms due to rising prices even as it diminishes in quantity terms. Thus the loss of these services due to global warming will have rising damage costs over time, and the rising cost of these damages offsets some or perhaps all of the effects of discounting. The result is a higher present value of damages and hence a stronger case for stringent controls on emissions. [21]

This general argument supporting rising relative value for environmental services is well known in the literature (Krutilla and Fisher 1975). The novelty in the global-warming context is that changes in relative prices are shown to affect the (endogenous) discount rate (Hoel and Sterner 2007). Discounting is treated more thoroughly in Chapter 3, but we can anticipate its link to the sustainability question. The critical parameter in the Hoel and Sterner exercise is the elasticity of substitution between environmental and conventional goods. If substitution is low, we can anticipate rising values for environmental services. The combined effects of rising relative damage costs and the associated adjustment in the discount rate are to produce a lower "effective" discount rate and perhaps a negative rate. This, of course, helps justify a more aggressive climate policy. Heal (2009) makes much the same point using consumption shares between environmental and conventional goods, rather than relative prices.

Neumayer (2007) constructs a potentially reinforcing argument. His basic point is that controversy over the correct discount rate is

---

[21] Alternatively, increasing difficulty is substituting man-made for declining natural capital (loss of biodiversity, declines in agricultural and fisheries productivity, etc.) may raise costs and reduce or possibly reverse traditionally measured economic growth.

misplaced. The real issue is the degree of substitutability of produced capital for natural capital (the strong versus weak sustainability rule) and thus the extent to which future climate change damages can be substituted with other consumption goods. If substitutability is strictly limited, it is not necessary to fuss about a "low" discount rate: "climate change will harm future generations in a way that no consumption growth, however high, can compensate for it" (p. 300). He argues that utilitarian-based BC analysis is still possible if the future damages to natural capital from temperature increase are properly estimated. But he also suggests an alternative: to scrap utility maximization as the sole objective and to make decisions with regard to future generations on a rights-based structure. In this approach, the need to maintain natural capital is treated as a constraint on our actions, necessary to respect the rights of future generations. Traditional BC, which maximizes discounted utility, is no longer a complete guide for policy.

A possible conflict with sustainability has led to other efforts at reframing traditional BC analysis. One approach is to *first* establishing an inter-generational distribution of rights to resources, including climate, on some acceptable moral basis, and *then* use the discount rate as a tool for establishing inter-temporally efficient investments (Norgaard and Howarth 1992). This inverts the normal BC sequence, which first looks at efficiency and then may concern itself with equity. Another approach is to seek a "green golden rule" in which some weight is given to present versus future consumption via the discount rate, but some additional weight is given to the welfare of distant generations (Beltratti, Chichilnisky, and Heal 1995). In this cobbled system, BC is retained but neither the present nor the future "dictate" the welfare of the other.

What are the rights that future generations should enjoy? Sustainability provides an easy answer. Sustainability has a clear meaning, at least to economists, as non-declining per-capita utility or real income. If we accept the idea that there is some critical level of natural capital (some critical change in global climate) the loss of which would violate sustainability, we then have a constrained maximization problem – maximize NPV subject to the constraint that utility does not fall. We are now moving away from classical BC toward what has been called the Tolerable Windows approach. The existence of uncertainty anywhere in the chain from emissions to economic impacts reinforces this shift.

Alternatives: Tolerable Windows, Safe Minimum Standards, Precautionary Approach, and the 2°C Target

Article 15 of the 1992 Rio Declaration on Environment and Development states: "In order to protect the environment, the precautionary approach shall be widely applied by States according to their capabilities. Where there is a threat of serious or irreversible damage, lack of full scientific certainty shall not be used as a reason for postponing cost-effective measures to prevent environmental degradation." As noted earlier, the UN Framework Convention on Climate Change, signed at the same Conference, has as its objective the stabilization of greenhouse gases at levels that would prevent dangerous interference with the climate system. Given the limits of conventional BC analysis discussed previously, does the precautionary principle offer a superior approach to global-warming policy?

Attempts have been made to translate the precautionary approach into specific policy guides (Petschel-Held et al. 1999). A basic premise is that climate variables should be kept within geologic ranges. This means global mean temperature kept below the last interglacial maximum of 16.1°C (plus a 0.5°C overage); the rate of temperature change, which is closely related to damages, kept to no more than 0.2°C per decade; and social/economic-reasons reductions in greenhouse gas emissions kept at moderate levels.[22] These constraints identified acceptable and unacceptable emissions trajectories, and thus what were considered safe minimum standards (a tolerable window), but they did not identify a single, efficient emissions path, nor, of course, did they demonstrate that benefits exceeded costs.

In the case of climate change, a precautionary approach would involve committing substantial resources to mitigating greenhouse gas emissions now to reduce the possibility of potentially severe but uncertain future damages. In itself, this is not necessarily in conflict with conventional BC analysis and its intellectual foundation, expected utility theory. If we use a very high, or a very very high, coefficient of

---

[22] The rates of change of concentrations and temperature change are important in determining damages. Weitzman (2009c) states that for the 800,000 years leading up to the industrial revolution, the rate of increase in CO_2 concentration was less than 25 parts per million (ppm) per thousand years, but reached 25 ppm in the last ten years alone.

risk aversion, we would have no difficulty in calculating a NPV that replicates a strongly precautionary approach. This, however, merely calls attention to our arbitrary choice of what is in essence an ethical parameter – how much do we wish to risk the welfare of future generations.[23]

In summary, modern BC analysis already folds in elements of the precautionary approach. Intelligent use of option values, the explicit modeling of climate in production and utility functions, and expected utility theory are three examples. The problem of relying on a pure precautionary policy is that it does not tell us how much we should spend or how fast. It would be seriously incomplete as a guide to policy.

A similar criticism can be leveled at the 2°C target, which has been widely adopted in policy declarations in recent years (European Council of the European Union 1996; Commission of the European Communities 2005; Copenhagen Accord 2009; Cancun Agreement 2010).[24] It is superficially attractive. By setting a target for allowable temperature increase, it appears that we can dispense with the messy and inclusive business of quantifying damages and finesse the ethical dilemmas of discounting over generations and social weighting of those damages. The problem simplifies to a least-cost exercise, always easier than a full-blown cost-benefit analysis. The 2°C target appears to have the added advantage of being based on science. For example, from the Copenhagen Accord: "... we shall, *recognizing the scientific view that the increase in global temperature should be below 2 degrees Celsius*, ... enhance our long-term cooperative efforts to combat climate change" (emphasis added).

In fact, both the simplicity and science basis for a 2°C target are an illusion. One cannot escape the need to consider the costs of such a policy and to set those costs against the expected benefits, namely damages avoided. The notion that a temperature target is based on science is also shaky. There is no consensus. For example, Ramanathan and Feng (2008) concluded that by 2005, countries had already emitted sufficient greenhouse gases to most likely commit the world to a

---

[23] The alternative – reducing the discount rate for precautionary actions that reduce future risks – goes against the conventional economic view that risk and time should be handled separately. The general public may find this injunction a bit pedantic.

[24] Tol (2007) is highly critical of the procedure through which the EU target has been set.

warming of 2.4°C on average, with a range of 1.4°C to 4.3°C, whereas Schellnhuber (2008) states there is "a fair chance" to hold the 2°C line but it will require an industrial revolution for sustainability now. In other words, we do not even know if the targe is technically feasible. Indeed, the IPCC has not endorsed this target. At a more fundamental level, to set a numerical temperature target without specifying the level of certainty required is not meaningful. Our policy tool is to control emissions. The steps between emission levels and temperature increases involve profound scientific uncertainties concerning the carbon cycle and the responsiveness of climate to concentrations. It follows that to translate targets into emission levels requires a temperature target and a number indicating what probability of success we can live with.[25]

Put somewhat differently, the 2°C conjures up a misleading image in which the curve relating temperature to damages has a hockey-stick appearance, with a sharp kink at 2°. It is an appealing image, suggesting a benign range of global warming, a tipping point, followed by serious consequences. No doubt there are tipping points but they remain largely unknown, and neither science nor economics lends great support to this approach at target setting. All this is perhaps unduly harsh. It may be useful in a political context to choose a nice round number as a focal point to rally support for strong action. Two degrees, if taken seriously, has very little slack. And we must remember the grave imperfections in the alternative approach, which is benefit cost.

Summary

BC analysis survives this litany of frailties but just barely. Its weaknesses include the dubious use of the KH hypothetical compensation test in an international and inter-generational context, the likely compounding of error through a long chain of scientific and economic assumptions, the inadequacy of its traditional tools for accommodating deep uncertainty and possible catastrophe, questionable ability to deal with thresholds, tipping points, and irreversibilities, and a neglect

---

[25] Meinshausen et al. (2009) provide a nice illustration. They calculate that a policy of limiting global emissions over the period 2000–2050 to 1,000 Gt CO_2 has a 75% probability of holding temperature increase below 2°C. If we could tolerate a lower probability of success of 50%, we could increase emissions to 1,440 Gt.

of sustainability concerns. In particular, the possibility of catastrophe tends to shift the debate away from conventional BC toward viewing policy through the lens of social insurance. The following chapter on discounting and social weighting adds to this list by documenting BC analysis's struggles with meeting intra- and inter-generational equity aspirations. Nevertheless, BC analysis and the structure on which it rest – expected utility theory – is a coherent framework for evaluating policy. It requires us to state our assumptions, including our attitude toward risk. Caution can be built in if we wish. It encourages a rational consideration of alternatives. And although it does not tell us what is equitable, it allows us to trace out some of the distributional implications of climate policy.

References

Arrow, K. and A. Fisher (1974). Environmental Preservation, Uncertainty, and Irreversibility. *Quarterly Journal of Economics* 88: 312–19.

Belli, P., J. Anderson, H. Barnum, J. Dixon, and J. Tan (2001). *Economic Analysis of Investment Operations*. Washington, DC: The World Bank.

Beltratti, A., G. Chichilnisky, and G. Heal (1995). Sustainable Growth and the Green Golden Rule. In *Economics of Sustainable Development*, Ian Golden and L. Alan Winters (eds.). Cambridge: Cambridge University Press.

Dietz, S. (2010). High Impact, Low Probability? An Empirical Analysis of Risk in the Economics of Climate Change. *Climatic Change*, published online December 16, 2011. Accessed at http://www.springerlink.com. proxyl.library.jhu.edu

Farrow, S. (1998). Environmental Equity and Sustainability: Rejecting the Kaldor-Hicks Criteria. *Ecological Economics* 27: 183–188.

Heal, G. (2009). Climate Economics: A Meta-Review and Some Suggestions for Future Research. *Review of Environmental Economics and Policy* 3 (1): 4–21.

Helm, C., T. Bruckner, and F. Toth (1999). Value Judgments and the Choice of Climate Protection Strategies. *International Journal of Social Economics* 26 (7/8/9): 974–98.

Hoel, M. and T. Sterner (2007). Discounting and Relative Prices: Assessing Future Environment al Damages. *Climatic Change* 84: 265–80.

Krutilla, J. and A. Fisher (1967). Conservation Reconsidered. *American Economic Review* 57 (4): 777–86.

 (1975). *The Economics of Natural Resources*. Baltimore: Johns Hopkins University Press.

Meadows, D. and D. Meadows (1972). *The Limits to Growth*. New York: Universe Books.

Meinshausen, M. et al. (2009). Greenhouse-gas Emission Targets for Limiting Global Warming to 2°C. *Nature* 458: 1158–63.

Mishan, E. (1988). *Cost Benefit Analysis: An Informal Introduction*. 4th ed. London: Unwin Hyman.

Neumayer, E. (2007). A Missed Opportunity: The Stern Report Fails to Tackle the Issue of Non-substitutable Natural Capital. *Global Environmental Change* 17: 297–301.

Norgaard, R. and R. Howarth (1992). Economics, Ethics and the Environment. In *Energy Environment Connection*, J. Hollander (ed.). Washington, DC: Island Press.

Petschel-Held, G., H-J. Schellnhuber, T. Brucknew, F. Toth, and K. Hasselman (1999). The Tolerable Windows Approach: Theoretical and Methodological Foundations. *Climatic Change* 41: 303–31.

Pindyck, R. (2007). Uncertainty in Environmental Economics. *Review of Environmental Economics and Policy* 1 (1): 45–65.

(2009). Uncertain Outcomes and Climate Change Policy. *NBER WP* 15259.

(2010). Fat Tails, Thin Tails, and Climate Policy. *NBER WP 16353*.

Rabl, A. and B. van der Zwaan (2009). Cost Benefit Analysis of Climate Change Dynamics: Uncertainties and the Value of Information. *Climatic Change* 96: 313–33.

Ramanathan, V. and Y. Feng (2008). On Avoiding Dangerous Anthropogenic Interference with the Climate System: Formidable Challenges Ahead. *PNAS* 105: 14245–50.

Schellnhuber, H. J. (2008). Global Warming: Stop Worrying, Start Panicking? *PNAS* 105 (38): 14239–40.

Sterner, T. and U. Martin Persson (2008). An Even Sterner Review: Introducing Relative Prices into the Discounting Debate. *Review of Environmental Economics and Policy* 2 (1): 61–76.

Tol, R. S. J. (2003). Is the Uncertainty About Climate Change Too Large for Expected Cost- Benefit Analysis? *Climatic Change* 56 (3): 265–89.

(2007). Europe's Long-term Climate target: A Critical Evaluation. *Energy Policy* 35: 424–32.

Tol, R. S. J. and G. Yohe (2007). Infinite Uncertainty, Forgotten Feedbacks and Cost-Benefit Analysis of Climate Policy. *Climatic Change* 83 (4): 429–42.

Viscusi, W. K. and J. Aldy (2003). The Value of Statistical Life: a Critical Review of Market Estimates throughout the World. *Journal of Risk and Uncertainty* 27 (1): 5–76.

von Bulow, D. and T. Persson (2008). Uncertainty, Climate Change, and the Global Economy. *NBER* WP 14426.

Weitzman, M. (2007). The Role of Uncertainty in the Economics of Catastrophic Climate Change. *AEI-Brookings Joint Center for Regulatory Studies* Working Paper 07–11.

(2009a). On Modeling and Interpreting the Economics of Catastrophic Climate Change. *Review of Economics and Statistics* 91 (1): 1–19.

(2009b). Some Basic Economics of Extreme Climate Change. *Mimeo* Feb 20.

(2009c). Reactions to the Nordhaus Critique. *Mimeo* March 17.

3

Discounting and Social Weighting
(Aggregating over Time and Space)

Introduction

This chapter continues our inquiry into whether benefit cost (BC) is an appropriate tool for setting global-warming policy. It considers the controversial and related issues of discounting and social weighting. Underlying the discounting debate is a central question: Given that damages will extend over centuries, does the very process of discounting inevitably rule out strong action today to limit global warming? Underlying the social weighting debate is another central question: If we discount the consumption of future generations because they may be richer than us, should we not also weight the losses to the poor more heavily than losses to the rich?

The structure of the chapter is as follows. After some preliminaries, we review the descriptive and prescriptive approaches to the discount rate. Next we examine the so-called Ramsey equation, which underpins most of the global-warming discount literature. The final section considers "social" or "equity" weighting of costs and benefits.

A word of encouragement is in order. Much of this is tough slogging, but the discount rate is of central importance. One reason the exposition is dense is that the underlying theoretical construct, which is based on discounted utilitarianism and the use of a mathematically convenient social utility function, is underspecified and relies on a single parameter (η) to play multiple roles. The result is a series of paradoxical results: A strong aversion to income inequality suggests a high discount rate, but a high discount rate works against a strong climate change program. It appears our egalitarian instincts conflict with our environmental stewardship responsibilities. Another example: a strong aversion to risk implies a high discount rate, but a high discount rate works in the direction of reducing mitigation efforts, increasing the possibility of climate catastrophe. How can a risk-averse society wind up courting runaway temperature increases? There is still another catch. If we are concerned about inter-generational equity, we also should be concerned about intra-generational equity. This is the question of the social weighting of global-warming damages, which will fall disproportionately on the poor. The same parameter, η, that plays a prominent role in the discussion of discounting reappears in a new guise, as a measure of our concern for intra-generational equality. In its discount role, the larger this parameter, the more minimal the mitigation effort, but in its new guise of promoting equality, the larger it is, the stronger is mitigation policy. Another conundrum. One way to wiggle out of these dilemmas is to reject traditional approaches and seek richer models. Another is to consider declining discount rates, which can be defended on theoretical and empirical grounds. There is much to sort out.

Discounting

If global-warming policy is to be approached through BC analysis, it is necessary to aggregate over time. For a meaningful comparison, costs and benefits must be summed to the same point in time. This could be somewhere in the future (compounding) but almost always is the present (discounting). The present value of costs and benefits that will occur in the future is thought to be less than the value they will have in the future. Thus their future values are multiplied by a discount factor, a positive number less than one, to obtain their present value. The discount *rate* is simply the rate at which the discount factor falls over time, and is expressed as a percentage.[1] It is also possible that the discount factor is a number greater than one, and increases over time. In that exceptional case, the discount rate is negative, implying that the present values of costs and benefits that will occur in the future are

---

[1] $DF_t = 1/(1 + r)^t$ where DF_t is the discount factor, r is the discount rate, and t is time. A constant discount rate means that the discount factor declines exponentially. It might help to think of the discount rate as an inter-temporal exchange rate through which values in one time period can be converted to values in another.

thought to be greater than the value they will have in the future. As discussed further in this chapter, it is also possible that the discount rate is not fixed but changes over time. This would solve some problems but create others.

The importance of discounting in global-warming economics cannot be overstated and is due to the very long time horizons. For example, $1 million of damages discounted over 300 years – the estimated average residence time in the atmosphere of a ton of carbon emitted today – has a present value of $15,439 when discounted at 1.4 percent, and less than $8 when discounted at 4 percent.[2] The former is the average discount rate used in the influential, if controversial, Stern Report. The latter is the average for the next 100 years used by William Nordhaus, a major figure in modeling climate policy, in his 2008 optimal climate policy exercise.

More generally, the discount rate is central to the conflated issues of efficiency and equity in the allocation of resources and welfare over generational time. Objections to discounting and the assertion that discounting discriminates against future generations are at the heart of many critiques of using BC in climate policy. The more primitive of these would weigh the present and the future equally, implying a zero discount rate. The more sophisticated critiques question the size of discount rate commonly used, and this is where the clash of views is sharpest.

Descriptive versus Prescriptive Approaches

Broadly speaking, there have been two approaches to establishing the discount rate. The first looks to the marginal rate of return on investment and is often called the descriptive approach. The second is the social rate of time preference (SRTP) and is called the prescriptive approach. The principal difference is that in the descriptive approach, discounting is based on observed market rates, whereas the social rate

---

[2] The residence time of CO_2 in the atmosphere and its measurement is contentious and confusing. See exchange of letters between Robert May, former President of the Royal Society, and Nobel Prizewinner Freeman Dayson, *New York Review of Books* 55(15) October 9, 2008. Archer and Brovkin (2008) state that "the largest fraction of CO_2 recovery will take place in timescales of centuries ... but a significant fraction of the fossil fuel CO_2, ranging from 20%–60% in published models, remains airborne for a thousand years or longer."

of time discount is built up from ethical criteria as well as behavioral evidence.

The distinction between the two approaches roughly mirrors a tension in the literature as to how closely the discount rate should reflect actual preferences and behavior, and how much it should reflect what we ought to be doing. Those who rely on evidence from financial markets emphasize the role of the marginal return on investment, the descriptive approach. But even those who are comfortable working directly with the prescriptive approach and the SRTP often attempt to ground their discount rates on evidence of observable behavior and preferences. For example, Pearce et al. (2003), among many others, state that SRTP is a *normative* construct, telling us what we should do. But he is also a strong supporter of respecting individuals' preferences as observed by what they actually do, and he searches for evidence to illuminate the SRTP. If preferences in fact count, revealed preferences help inform what ought to be.

The choice between using and not using market information is not as stark as it might appear. It is possible to use the SRTP but also capture the opportunity cost of private investments that might be displaced by global-warming mitigation measures. This involves calculating the "shadow price" of capital in terms of its consumption equivalence. Consider, for example, a tax-distorted economy in which the pre-tax marginal return on investment is higher than the after-tax return earned by savers. The first is a measure of the productivity of capital and the second derives from society's preferences for present versus future consumption and the premium it expects for deferring consumption (i.e., for saving). The two rates are equal under idealized conditions, but the existence of taxes and other distortions drives a wedge between them, with the return on investment greater than the return to savers. The two rates can be reconciled using the notion of the shadow price of capital, which converts investment flows into their consumption equivalent values and then discounts all flows – consumption flows and the consumption equivalent of investment flows – at the SRTP.[3] In effect, this weighs a unit of investment more highly

---

[3] The tricky part is to determine the fraction of project financing diverted from private investment and subject to shadow pricing, and the fraction diverted from consumption. Indeed, it is necessary to estimate all subsequent flows consequent to the project into and out of consumption and reinvestment. The shadow price of capital should,

than a unit of consumption. It is easy to become confused at this point because some older manuals for developing countries would shadow price *consumption* flows to determine their government income value, and then discount all flows at the accounting rate of interest, which is the rate of fall of the value of a unit of government income over time. This procedure illustrates the (then) prevalent but now dated view that a unit of resources in the hands of the government should be given a premium in development.

Neither the descriptive nor the prescriptive approach is a clear trump on ethical grounds. The SRTP approach leads to lower discount rates and thus justifies greater expenditures to protect future generations from global-warming damages. However, as we shall see, it relies on questionable assumptions about two value parameters – the so-called pure rate of time preference and the responsiveness of utility (well-being) to changes in consumption. Supporters of using higher rates based on marginal return on investment argue that our obligation to future generations is to invest in those projects and policies that yield the highest return. If investments in global-warming mitigation cannot compete with investments in health, or education, or infrastructure, or general research and development (R&D), we would shortchange the future if we used an artificially low discount rate for climate measures. If our concern is for the current and future poor, foreign aid today might yield higher returns than very-long-term investment in preventing climate change.

This is a strong argument. However, it also points to a major weakness in the descriptive approach. This weakness is the inability to creditably commit to transfer resources over many generations as described in Chapter 2.[4] There is no way of guaranteeing that resources not spent on mitigation today will be invested in a "sinking fund" and invested and reinvested over many generations to compensate those who will be harmed. The Kaldor–Hicks hypothetical compensation test is compromised.

---

in principle, also reflect externalities, including global warming itself. See Boardman et al. (2006).

[4] Of course, intermediate generations could abort a strong abatement program as well. But there is a difference between undertaking activities today that are known to cause harm in the future, and hoping that society will accumulate and transfer forward in time sufficient resources to compensate for the harm, and avoiding the harm today and hoping that intermediate generations will do the same.

There are other reasons for questioning the use of market rates. These rates may be seriously distorted from environmental externalities including global warming itself. Specifically, the market rate of return to capital tends to be inflated as it does not measure the depletion of natural capital, including the atmosphere. Also it is unclear which rate of return to capital should be used, as there are many market rates.

This descriptive-prescriptive debate on discounting has become somewhat stale over the years.[5] Perhaps more importantly, new developments in discounting research, including the case for declining discount rates and the interaction between uncertainty and discount rates, have shifted the terms of the controversy. Even so, discounting continues to play a central role in climate economics. Because the debate still smolders, it is worth our effort to understand it.

The Ramsey Equation

Both the descriptive and prescriptive approaches to discounting make use of the Ramsey equation, but in quite different ways. In an optimal-growth model, in which the objective is to maximize the discounted utility of consumption, Ramsey (1928) showed that the social rate of time preference (SRTP) is equal to the sum of two terms, a "pure rate of time preference", ρ, and a second term that multiplies the expected growth in per-capita consumption, c, by a weighting parameter, η. He also showed that in an undistorted economy, the SRTP was equal to the marginal rate of return on investment, which we can call r. Thus,

$$SRTP = \rho + \eta c = r.$$

The basic distinction between the descriptive and prescriptive approaches is that the descriptive approach calibrates the values of ρ and η to historical data on the market rate of return on investment,

---

[5] It was featured in the World Bank publication, *Finance and Development*, in March 1993, with articles by William Cline (prescriptive) and Nancy Birdsall and Andrew Steers (descriptive); was thoroughly addressed in the IPCC 1995 Second Assessment Report, was reconsidered by the experts in an excellent conference volume edited by Portney and Weyant (1999) and sponsored by Resources for the Future, and was debated by Cline, Mendelshon, and Manne in the 2004 Copenhagen Consensus exercise. Most recently, two prominent proponents, Nicholas Stern (prescriptive) and William Nordhaus (descriptive), reprised the arguments in their recent books (2007, 2008), but without a clear knockout by either.

whereas the prescriptive approach seeks ethically and empirically reasonable values for these two parameters and pays little attention to
market rates of return. We can immediately spot potential difficulties.
The descriptive approach does not adduce any evidence for ρ and η
but does require their size to be consistent with the projected rate
of return on investment. Given projected growth of consumption, the
values for ρ and η are "backed out" from market return evidence. The
allocation as between ρ and η tends to be arbitrary, although it makes
a difference in the long run. Moreover, useful information about the
appropriate *social* rate of discount – a normative concept that reflects
our ethical responsibilities to future generations – is difficult to infer
from the private behavior of individuals in financial markets.[6]

The prescriptive approach lacks any definitive source for its two
normative parameters, ρ and η, and risks coming up with values that,
in combination with the expected growth of consumption, falls short
of the observed rate of return on investment. As explained later in the
chapter, implausibly high savings rates would be implied. Those most
concerned – future generations – of course are not consulted in either
approach, for the obvious reasons.

The Ramsey equation invites closer scrutiny but before doing so we
make one initial comment. The equation is firmly centered in the tradi-
tion of maximizing discounted utility when considering inter-temporal
allocation decisions. This is common in the literature, but there are
alternative ethical approaches that yield different inter-temporal welfare allocations. The stewardship approach holds that each generation
bestows on the next at least as much productive capital (environmental and other) as it enjoys; the Rawlsian approach maximizes the welfare of the poorest generation; the precautionary approach attempts
to minimize inter-temporal risk.

Rho, the Pure Time Preference Parameter

The Ramsey equation, as used in the global-warming literature, considers and compares welfare among generations. It does not purport to

---

[6] Even if individuals recognized that financial markets were a vehicle for expressing
inter-generational preferences, a market failure may be present. The isolation paradox
holds that saving for the future has something of the character of a public good, and
may be undersubscribed without some collective commitment. Thus market rates may
diverge from social rates.

describe an individual's preferences. Nordhaus (2008, p. 172) is quite explicit about this: "[T]he variables analyzed here apply to the welfare of different generations and not to individual preferences. The individual rate of time preference, risk preference, and utility function do not, in principle at least, enter into the discussion or arguments at all." This is important because at the level of an individual, there are theoretical and empirical reasons for believing that ρ may be positive and substantial, but these reasons lose force when considering inter-generational welfare, the hallmark of global warming. Specifically, individuals may exhibit impatience and pay a large premium for present consumption (for example, at credit card rates). But our impatience is for our own consumption. Inter-generational societal impatience makes no sense. We may feel somewhat stronger *affinity* for a generation 100 years from now as compared to a generation 1,000 years from now, but this is emotional distance, not impatience for our consumption. Upon reflection, impatience is an individual characteristic and not relevant to inter-generational discounting.

The other individualistic rationale for a positive ρ – our mortality and our realization that we may not be around to enjoy delayed consumption – is weakened. We can be assured of our individual mortality. Society is another matter, however. There is certainly a *possibility* of human extinction. Stern (2007) sets it at the remarkably pessimistic high probability of 10 percent over the next 100 years, which translates into ρ equal to 0.1 percent. However, this Armageddon outlook for society is only weakly connected to the prospect of our own personal demise. One cannot infer much useful information about the fate of humanity from the mortality of individuals.

The choice of ρ, the pure time preference, thus appears to be an ethical one. Should we discriminate against future generations by using a positive ρ simply because they will live in the future (and coincidentally are not here to protest)? This is a question of values and cannot be given a definitive answer. Many respected economists have argued on ethical grounds against inter-temporal discrimination (e.g., Ramsey himself, Pigou, Harrod, Solow, Dasgupta, Heal).[7] Other considerations are also relevant, however. Setting ρ equal to zero implies that we care

---

[7] Dasgupta (2001) quotes Ramsey on discounting the well-being of future generations as "ethically indefensible and arises merely from the weakness of the imagination."

as much about the welfare of generations hundreds and even thousands of years from now as we care for our own welfare. This seems to defy evidence and common sense. Remember, we are now abstracting from any income changes.

A more complicated argument is that the Ramsey equation does not give us a completely free hand to select ρ. In the underlying-growth model, a zero value for pure time preference would imply an implausibly high current savings rate approaching 100 percent, unless offset by a very high value given to η.[8] Koopmans (1967) has called this "the paradox of the indefinitely postponed splurge." The intuitive reason is that current resources devoted to savings and investment would earn positive rates of return forever, the present value of which, discounted at a zero pure rate of time discount, would exceed any loss of welfare to the current generation from its heroic self-sacrifice. It is true that the disturbing prospect of our sacrifices – our impoverishment – for the sake of the future could be offset if we chose a sufficiently large η, as some have suggested. However, as we will see, the several roles η plays are complicated enough without yoking it to ρ just to avoid an implausibly high savings requirement.

The problem of excessive savings may be overstated. The equality of the SRTP and the marginal rate of return on investment only holds in an ideal economy with no distortions. But global warming has been called the mother of all externalities. It may be that when externalities are accounted for, the *social* rate of return to investment is zero or negative, and the entire excess-savings issue evaporates. Moreover, it has been argued that adding technical progress to this type of model can sharply reduce implied savings rates.[9] Recall also that the assumed constancy of η is only for analytic convenience, and if it were allowed to vary over time, the excessive savings issue loses traction.

A straightforward and honest approach to ρ might be to acknowledge that we do have inter-temporal preferences regarding the welfare of future generations, but that these have little to do with impatience for our own consumption, or our own mortality. It would not seem unethical to admit that the welfare of our immediate successor generations have a higher value than generations that emerge in

---

[8] For an example, see Dasgupta (2001), Appendix A13.
[9] Dietz and Stern (2008).

the far future. We could then dispense with the fiction that all generations are equally valuable and concentrate on thinking about the rate
at which the utility discount factor for future generations (ρ) should
decline. It would also allow that rate to vary over time, and we could
contemplate declining social discount rates. While such an admission
would be a retreat from strict utilitarianism, it would help maintain
a link between the SRTD and what is likely to be our actual time
preferences.

Adjusting for Consumption Growth

We now consider the second term in the Ramsey equation, ηc. The purpose of this term is to adjust the SRTD for likely increases in consumption and a related *assumption* that marginal utility (MU) declines with
higher consumption.[10] We take this in two steps, first c and then η.

The expected rate of growth of per-capita consumption, c, is not
a question of values. It can be treated as exogenous or endogenous
in the analysis but is presumably free of ethical considerations. It is
not a choice parameter. The recent recognition of environmental services as inputs into both production functions and utility functions
has two implications for c. First, conventionally measured national
income statistics most likely overestimate the rate of growth of real
consumption, including non-marketed environmental services, as neither the damages of pollution nor the depletion of natural capital is
adequately captured. Second, spending on environmental protection
tends to reduce the productivity of capital as conventionally measured
(Weitzman 1994). Both of these have the effect of lowering c and thus
the SRTP. *Ceteris paribus*, these adjustments support a more aggressive greenhouse gas abatement program.

A more radical adjustment is to consider c as both endogenous and
uncertain in modeling global warming. Some, not all, studies drop the
assumption that global climate change is necessarily "small" relative
to our economic future. Whereas BC analysis has traditionally used
the marginal assumption to justify partial equilibrium analysis, this
may not do justice to the magnitude of global-warming effects. It

---

[10] The following perhaps tedious discussion could be short-circuited if the richer future
would compensate us, the poor present, for our abatement expense. They cannot and
they will not.

follows that for proper analysis of an emission mitigation policy it may be necessary to contemplate an array of consumption paths, each with its own probability. If an acceptable probability distribution is available, the expected value of c as a probability-weighted consumption path can then be calculated.

Although c – the expected rate of growth of per-capita consumption – is not a value choice, it is critical to get it right. Together with ρ and η, it helps determine the social rate of time discount, and hence how we treat future generations. If we underestimate the negative impact of global warming on productive capacity, and thereby overestimate future consumption levels, we shortchange the future, and the minimum sustainability criterion of non-declining income is put in jeopardy.

Deconstructing Eta

We have circled around η and now must confront it directly. *The central meaning of η is a parameter that describes the strength of diminishing marginal utility of consumption.* For the more visually minded, it is the measure of the curvature of the utility function. The higher η, the more sharply MU declines. The higher η, the less *additional* satisfaction from the third glass of wine, but still a positive experience.

As a measure of the strength of diminishing marginal utility, η plays three roles in global-warming economics. First, it is put to work in the Ramsay discounting equation because we expect future generations to be wealthier than ours and we need to know how much utility to attach to those higher levels of consumption. Second, because climate change involves risk and uncertainty, we need a measure of how averse we are to risk. As explained later, risk aversion is related to declining MU and hence is conveniently measured by η. Third, cost-benefit analysis of global warming requires aggregating the effects on individuals and countries with greatly different incomes. Social weights can be constructed to adjust for differences in the utility or well-being associated with consumption by different income groups. The weights are based on the rate at which MU is thought to decline, and this is the third role η plays.

It may be helpful to be more explicit. Eta's origins are from an assumed social utility function that exhibits diminishing MU of consumption. The most popular but by no means the only candidate is the so-called iso-elastic function,

$$U(C) = (1/1-\eta)C^{1-\eta} \quad \text{for } \eta>0, \eta \neq 1$$
$$\text{And } U(C) = \ln C \quad \text{for } \eta = 1.$$

This is mathematically convenient, as MU, the increment in utility from an additional unit of consumption, is then

$$MU = 1/C^{\eta},$$

and (minus) η is the elasticity of MU with respect to consumption, the percent by which MU declines with a 1 percent increase in consumption. The iso-elastic simply means that with this particular *assumed* utility function, the elasticity stays the same no matter what the consumption level, although there is no obvious reason for making this assumption.

Eta is a powerful parameter. It is worthwhile putting some numbers on it to see the implications. If η were zero, there would be no diminishing MU of consumption. If η were 1, and if we normalize on you, the reader, a dollar of income to someone at half your income level would be valued at 2 dollars, and a dollar to someone at twice your income level would be valued at 0.5 dollars. If η were judged to be 2, a number often used at least for illustration, a dollar taken from someone who is relatively rich (ten times average income) and transferred to someone who is relatively poor (one-tenth the average income) should be valued at $10,000.

Assuming income grows, it is difficult to overestimate the importance of η in discounting damages from global warming that occur 100 or 200 years from now. Consider, for example, the present value of $1 million in global-warming damages that occur 100 years from now. To stay focused on η, assume ρ is zero (no discrimination based purely on time). If per-capita income grows at 2 percent per year and we assume η equals 1, the present value of the damages is $138,000. If η equals 2, the present value of the $1 million is $20,000, and for η at 3 – a value supported by some economists – the value drops below $3,000. While ρ, the pure rate of time preference, has received the most attention in the literature, η is of equal importance. It is more complex than ρ, however, serving multiple roles.

Consider the three roles of η. When employed in the Ramsey equation, η is the weight used to find the present value of future increases in per-capita consumption. It reflects social choices about equality and inequality in the distribution of income across generations. It is a measure of our aversion to inter-temporal inequality in consumption or, in more positive terms, it is a measure of our inter-temporal egalitarian preferences. Our preferences are undoubtedly influenced by the concept of diminishing marginal utility. If future generations are richer than ours, the satisfaction or utility they obtain from a dollar of additional consumption will be less than we would receive. But our aversion to inter-temporal income inequality is also grounded in some notion of fairness in the distribution of consumption across generations, a notion that is independent of diminishing marginal utility. How much should the current "poor" generation sacrifice for "rich" future generations? Declining MU and fairness both matter in the choice of η.

We must be careful to avoid confusion. The pure rate of time preference, ρ, and η tend to work in the same direction with regard to global warming. The lower their values, the lower the discount rate, the higher the present value of global-warming damages, and the stronger should be the abatement effort.[11] They do *not*, however, work together to simultaneously support a strong abatement effort and inter-generational equality. On the contrary, inter-generational equality demands a large η, whereas the abatement objective needs a low ρ and a low η. The dilemma mentioned in the introduction to this chapter is real. Our environmental stewardship responsibilities conflict with our egalitarian instincts. This is an important point. A strong preference for inter-generational equality does *not* protect the environmental patrimony of future generations. Instead it protects the interests of the current "poor" generation from excessive subsidization of (presumably) rich future generations. Aversion to inter-temporal inequality redresses consumption in our favor and in no sense husbands resources for the future.

---

[11] This is correct only if consumption growth is positive. If it is negative, a low ρ continues to work for a strong abatement effort, but the lower η, the higher the discount rate. Note also that if ρ is sufficiently low and ηc is sufficiently negative, the discount rate itself is negative.

Now consider the second role η plays. It is the coefficient of relative risk aversion in the finance literature, measuring the extent to which risk-averse individuals accept lower investment returns in exchange for lower risk. It measures the convexity of the utility function, the same role it plays in the Ramsey discount equation. The larger η, the more convex is the function and the greater the aversion to risk. Moreover, using the same iso-elastic function introduced earlier produces a constant relative risk-aversion coefficient. This implies that people at different consumption levels would pay the same fraction of their consumption to avoid a given consumption loss. The rich and the poor are assumed to avoid risk in the same proportion.

The future state of the world is uncertain and is reflected in uncertain levels of future consumption. A strong preference for risk aversion implies a large η and, in an expected utility framework, places most weight on the worst consumption outcomes. This increases the expected value of global-warming damages and supports a strong climate policy. But we see another paradox here. A strong aversion to risk calls for a high η, which increases the future value of damages. But a high η, working through the Ramsey equation, will reduce the present value of these damages, weaken climate policy, and increase the possibility of runaway temperature increases and catastrophe. The net effect is ambiguous. The fundamental problem is that too much is being asked of the single parameter, η.[12]

As if this were not enough, η plays a third role. Not only is it necessary to aggregate over time, but also over countries at different income levels. The utilitarian social welfare function aggregates *utilities* and that means converting consumption levels into utility presumably using the same utility function as used in aggregating over time. In this third role, η is a measure of *intra-generational* inequality aversion, or our preference for an equalitarian distribution of income at a point in time. The larger η, the stronger is our aversion to inequality. In this role, η fixes social or equity weights, and is discussed in more detail later. For the moment, it is only necessary to note two things. First, our aversion to intra-generational inequality has the same two wellsprings

---

[12] Saelen et al. (2008) stress these conflicting effects and refer to work by Simon Dietz who found a U-shape for the present value of damages as η is increased. This is consistent with Weitzman (2010), who finds that with "fat tailed" probability distributions, the risk aversion role of η comes to dominate its discount rate role.

as our aversion to inter-generational inequality: respect for the idea of diminishing MU *and* a sense of fairness. The loss of a dollar to a poor person should weigh more heavily than the same value of loss to a rich person. Fairness becomes important because, as discussed in Chapter 2, the current and prospective international distribution of income is manifestly unfair, with no obvious political mechanism for correction. Second, note that there is no compelling reason to believe that our preferences for equality in consumption across generations are identical to our preferences for equality of consumption as between countries (or individuals) at a point in time.

Now we see yet another paradox. It is widely agreed that global warming will disproportionately harm poor countries and especially the poor in those countries. If social weighting is used, those damages would be given a higher weight, and *ceteris paribus*, greater damages would justify a *strong* climate policy today.[13] A strong aversion to intra-generational inequality suggests a large η, which means larger damages. But as we have seen, acting through its role in the Ramsey equation, a large η also reduces the present value of future damages. It would then help justify a *weak* climate policy. Once again, the net effect is ambiguous. A strong aversion to intra-generational inequality could contribute to *increasing* total future damages, but via discounting reduce their present value, and fail to protect the poor.

Estimating Eta

Despite its importance, there is no consensus about the appropriate value of η. This is not surprising. The three roles suggest that three different numbers may in fact be correct. Nordhaus uses a value for η of 1, as does Stern. Pearce and Ulph (1999) settled on 0.8. Cowell and Gardiner (1999) give 0.5–4 as the appropriate range. Dasgupta (2008), relying on the excessive-savings argument, feels comfortable with a range of 2 to 3 and perhaps higher. Weitzman (2010) uses 3 in his base case. This is not a trivial range for one of the most powerful parameters in climate economics.

One difficulty is that we have almost no direct evidence on society's aversion to inter-generational income inequality, η's first role.

---

[13] In an early study, Fankhauser et al. (1997) found that damages would be three-to-five times higher if social weights based on η equal to 2 were used.

Eventually global-warming policy may reveal something about those preferences, but that does not help in framing the policy. Evidence from finance and risk markets is not much help. Market returns on investment and implied aversion to risk reflect individuals' concern for their own and perhaps their children's generation, not a concern for inter-generational income inequality. Even then the rate of return is arbitrarily allocated between ρ and η in the descriptive approach.

In the absence of direct evidence, values for η are sometimes inferred from *intra*-generational income inequality aversion data, mainly progressive tax structures in advanced democratic countries. Evans and Sezar (2005), for example, infer from European tax structures that η ranges from 1.1 for Sweden to 1.8 for Luxembourg. These exercises do not instill great confidence in BC analysis of global-warming policy. It is questionable to use a numerical value of η derived from, say, the U.S. or European tax structures to put a present value on lives or crops lost from increased flooding in the Mekong delta 100 years from now.

Inter- and intra-generational aversion to inequality may also be very different. This is an important point. We have obligations to future generations. We also have obligations to the poor among us today. However, the two should not be conflated. In particular, the rich countries are primarily responsible for the global-warming problem. This creates obligations toward future generations that may moderate or outweigh the prospect that they will be richer than we. In contrast, the rich among us today are not directly responsible for the poverty of the many (with some obvious exceptions). We have obligations to the poor today, not because we are primarily responsible, but because it is the ethical thing to do.[14] The contexts in which inter- and intra-generational inequality preferences are formed are different. Data from foreign aid as a proxy for intra-generational preferences are suspect. At best they reflect the egalitarian preferences of rich donors, but not the poor.

These difficulties in pinning down η, and indeed the multiple roles for η, are illustrated by a survey conducted by Atkinson and his co-authors (2009). The survey shows that individuals' attitudes toward risk, their aversion to inequality at a point in time, and their aversion

---

[14] It would be odd if our obligations to those who are present and with us today elicit the same feelings as those who are not here and who will live in the distant future.

to inter-generational inequality are only very weakly correlated. The standard theory groups these dimensions into the single parameter, η, forcing a perfect correlation. Moreover, the survey reveals it is not uncommon for individuals to hold "conflicting" preferences, for example high risk aversion but little concern for income inequality. Such preferences cannot be accommodated within a single value for η.

Taking Stock

The multiple roles played by η and its numerical indeterminacy create problems for designing climate policy. The social cost of carbon is a widely used measure of the stringency of mitigation policy. It is the present (discounted) value of the stream of damages that would result from emitting one more ton of carbon. It has direct policy significance and provides a first approximation to the tax that should be placed on carbon emissions. If estimated through an optimization model, it measures the optimal (Pigouvian) tax. The reliability of BC as a guide to policy relies in large part on the accuracy of estimating the social costs. But the effect of η on the social cost of carbon is itself indeterminate. Hope (2008) has calculated that increasing η will lead to a *decrease* in the social cost of carbon because the increases in the discount rate *a la Ramsey* will more than offset the social weighted increase in damage costs. The interactions of uncertain values for pure time preference, ρ, and for η in its various roles, can also yield greatly different estimates of the social cost of carbon. This also erodes the reliability of BC as a guide to policy. Anthoff, Tol, and Yohe (2009a) use an integrated assessment model (FUND) to calculate this cost over a range of values for these parameters. The exercise compares results with and without risk aversion, and with and without social weighting of damages, and thus helps disentangle the three roles of η. The social cost of carbon ranges from *minus* $50/t carbon ($\rho = 3, \eta = 3$, social weighting included but no risk aversion) to $152,155/t carbon ($\rho = 0, \eta = 3$, risk aversion included, social weighting not included).[15] The authors have narrowed this down to a final estimate of the expected social cost of carbon,

---

[15] The negative social cost means a positive benefit and apparently arises from improved agricultural yields in the near future, heavily weighted in developing countries. Long-term losses are very heavily discounted through high values for ρ and η. In subsequent research (Anthoff, Tol, and Yohe 2009b), the authors also experiment with differing impact elasticities by sector and region.

including uncertainty and social weighting, of $206/t. The two parameter values they use for this estimate are based in part on the work of Evans and Sezer for European countries, who used national mortality rates as the basis for estimates of ρ and tax structures (inequality aversion) for estimates of η. The range of estimates and the sources of best estimates do not inspire confidence in BC.

It is time to take stock. The most commonly used theoretical structure assigns multiple and potentially conflicting roles to a critical parameter, η. This is further complicated by the flimsy empirical basis for both η and the pure rate of time preference, ρ. We will now do what we can to work our way out of this muddle and then return to the issue of social weighting.

Unsnarling the Discount Rate Tangle

Richer Models

Much of the problem arises from the triple role of η. This suggests that the basic model consisting of a utilitarian social welfare function and an iso-elastic utility function is underspecified. There have been some attempts to disentangle risk aversion, inter-generational inequality aversion, and intra-generational inequality aversion, but there is no model that disentangles all three concepts simultaneously. A start has been made, however. Ha-Duong and Treich (2004) build a model that contains separate parameters for risk aversion and intertemporal inequality aversion. They find that the two work in opposite directions with regard to optimal mitigation effort. High risk aversion leads to a stronger mitigation effort, whereas a strong aversion to inter-generational inequality works to reduce mitigation effort. When these two parameters are bundled into a single number, η, as they are in the standard treatment, the separate effects of risk aversion and inequality aversion preferences are obscured. The full impact of risk aversion preferences on mitigation effort is muffled.

Recalculating Damages

Another tactic is to shift attention from the discount rate to the global-warming damages that are being discounted. Chapter 2 suggested that in a two-sector model in which environmental services become increasingly scarce, the loss of these services due to global warming will inflict rising damage costs over time. Increasing damage costs offsets some or all of the effects of high discount rates. If, for example, damage costs are rising at 3 percent per year and are being discounted at 4 percent, the *effective* discount rate is only 1 percent. If the substitutability of produced and natural capital is limited, increasing damage costs can overwhelm even high discount rates, and aggressive action to protect climate for future generations may be justified. Hoel and Sterner (2007) construct a two-sector model (environmental good, conventional good) within a traditional inter-temporal utility maximization framework. They show that the lower the growth rate of the environmental good, and the lower the elasticity of substitution between the environmental and conventional good, the lower will be the *effective* discount rate – the discount rate minus the rate at which environmental damages are increasing. Not only is the effective discount rate below the rate in a one-sector model, but over time, both the discount rate and the effective discount rate decline as the environmental good becomes increasingly scarce. Provided the elasticity of substitution is limited, the application of this approach will reduce the power of discounting. It can go some distance to countering opposition to discounting and the use of BC analysis to inform global-warming policy.

Declining Discount Rates

Declining discount rates, sometimes known as hyperbolic rates, appear to offer some easing of the dilemmas the Ramsey equation reveals. Until very recently, cost benefit assumed that the discount rate, once selected, would remain fixed over the life of a project. There is no theoretical justification for this assumption; it is simply a convenience. It works well for small projects and in the short and medium term, but less so for inter-generational problems such as climate change. Moreover there is considerable research showing that people behave as though discount rates are not fixed over time. Specifically, individuals tend to discount over the near future at a higher rate than in the distant future. The premium one expects from deferring consumption from this year to the next is higher than the premium expected for deferring consumption from ten years to eleven years hence.

Following the Ramsey equation, declining discount rates will emerge in a deterministic setting if income and consumption growth slows and ρ and η are held constant.[16] Long-run growth is speculative and declining rates are not implausible. One explanation for slower growth is that an increasing fraction of investment may have to be spent on environmental protection, especially if the marginal productivity of that investment in that activity is declining. Global warming itself is expected to have a negative effect on productivity.

Uncertainty about future discount rates provides an additional, theoretical justification for rates that decline over time (Weitzman 1998). The discount rate is nothing more than the rate at which the discount factor declines over time. If it declines at a slower rate in the far future than it does in the near future, the discount rate appropriate for the far future is lower than the discount rate for the near future – that is, declining discount rates. Why would the decline in the discount factor slow down? Assume that the future discount rate is uncertain, with a one-third probability that it is 1 percent, a one-third probability of 3 percent, and a one-third probability of 5 percent.[17] In the very near term, the certainty equivalent discount factor (the risk-adjusted average discount factor) is close to 0.97 and the initial discount rate is 3 percent.[18] Over time, the weights of the higher discount rates in determining the discount factor decline, the certainty equivalent discount factor falls at a slower rate, and the certainty equivalent discount rate falls. After 200 years, the latter is 1.55 percent and will approach the lowest possible rate, 1 percent, in the very long term. The theoretical result of declining discount rates requires that there be some persistence in the discount rate itself, and empirical work on past uncertainty in U.S. interest rates tend to confirm this (Newell and Pizer 2003).

Declining discount rates, which moderate the harsh intergenerational implications of global warming, are not without their own problems. Time inconsistency is the main one. It occurs when an

---

[16] Models that calculate regional economic growth can determine region-specific discount rates.

[17] We borrow the numerical example from Guo et al. (2006).

[18] Using the equation $CEDF_t = 1/(1+CEDR_t)$ where $CEDF_t$ is the certainty equivalent discount factor at time t, and $CEDR_t$ is the certainty equivalent discount rate at t. In year 1, $CEDF = 1/3[1/1.01+1/1.03+1/1.05] = 0.971$.

action that appears optimal at one date no longer appears optimal at a later date. Loosely speaking, if today it appeared optimal to delay consumption from year fifty to year fifty-one in exchange for a 2 percent bonus, when year fifty rolls around and the then current bonuses are 6 percent, it no longer looks like a good deal. What looked optimal is then suboptimal. If this can be anticipated, as it could with time-declining rates for global warming, it leaves open the question of the appropriate action today. Perhaps too much should not be made of time inconsistency in the climate context. By the time the inconsistency materializes, the knowledge base for making policy will be far advanced, and policy shifts will be inevitable. And the time inconsistency problem only arises when the utility discount rate, ρ, changes over time, and not when the other component associated with consumption, ηc, changes (Hepburn 2006).

The extent to which declining discount rates might affect policy analysis depends on the model. Groom et al. (2005) conclude that the estimates of the social cost of carbon – the most common indicator of the optimum level of abatement – are likely to at least double if declining discount rates are employed. Guo et al. (2006) find that depending on the discounting scheme analyzed, declining discount rates increase the social cost of carbon from 10 percent up to a factor of 40. Declining rates may ease the concern that discounting discriminates against future generations, but they do not provide an easy answer to what specific rates or rates of decline we should be using.

Social (Equity) Weighting: Aggregating over Space

Concepts

We now turn to a more complete look at social weighting, which increases the value of costs and benefits incurred by the poor and decreases the values incurred by the rich. Why bother to do this? Unlike weighting consumption between rich and poor generations, which is central to all discounting exercises and is never questioned, social weighting of rich and poor within a generation is often treated as optional in global-warming literature. The Intergovernmental Panel on Climate Change (IPCC) has shown how social weighting of global-warming damages could be done, but has not followed through.[19] The Stern Report discussed social weighting in detail and favorably, but did not use this approach, citing time constraints.[20] It did, however, speculate that social weighting might increase the damages of climate change by 25 percent. Many other eminent IAM modelers have simply avoided social weighting. In fact, social weighting in global-warming analysis has been the exception rather than the rule.[21] This is troubling.

The answer to our question – why weight? – is the same for intra-generational income disparities as it was for inter-generational income disparities when considering the Ramsey equation. In both cases, it rests on an aversion to income inequality and is founded on a belief in diminishing MU of consumption and an ethical sense of fairness. In most cost-benefit analyses, social weighting can be finessed because they are done at the national or subnational level. As explained in Chapter 2, governments have the authority and tools to achieve a fair distribution of income and we can be reasonably comfortable about neglecting distribution, and thus social weighting. This does not wash for global warming, however. The international distribution of income is widely considered unfair, and is likely to remain unfair, and there are no credible institutions for compensating those damaged by global warming. It is therefore especially important that the income distributional incidence of damages is correctly estimated and valued, which is what social weighting attempts to do.[22]

As noted previously, all studies conclude that global warming will harm poor countries disproportionately.[23] The reasons are clear. The

[19] See especially IPCC Working Group III, chapter 6, SAR, and the efforts by Fankhauser et al. (1997) to show how weighting can change the monetary value of damages. Without explicit weights, the default weight is 1. A dollar to the rich and a dollar to the poor are treated equally. Kverndokk and Rose (2008) consider the use of equity weights in a broader discussion of ethics and justice.

[20] Stern Report, p. 282.

[21] This is beginning to change. For example, Anthoff, Hepburn, and Tol (2009) find that using regional welfare functions, rather than the global average function used by Stern, results in much larger impacts than Stern speculated.

[22] A bit of history is relevant. Social weighting was promoted in project evaluation in the 1970s at the World Bank and elsewhere, but fell out of favor in the next decade. The justification for weighting at the time was that governments in developing countries had very limited tools to achieve distributional objectives, and biasing project selection to favor the poor through social weighting was the second-best alternative.

[23] One study focusing only on market damages estimated that in 2100, the poorest 25 percent of world population would suffer damages from 12 to 24 percent of their

structure of their economies is weighted toward agriculture and natural resources, which are expected to be hardest hit. Poor countries have limited resources and capacity for adaptation. And they are primarily located in the tropics, where temperatures are already above the optimum. The case for social weighting does not, however, rest on the disproportionate impacts on the poor. The case rests on persistent differences in the MU of income between rich and poor countries. Currently poor countries will not necessarily remain poor. It is the future distribution of income, not the current distribution, that is at issue. If income levels converge rapidly, the case for social weighting is weaker. But even with more rapid growth in poor countries, substantial differences in per-capita income are expected to continue well into the twenty-second century. For example, if India's and U.S. per-capita income grew at 3 and 1 percent per year, respectively for the next 100 years, India's level would still be only 42 percent of the U.S. level.[24]

Consistency provides another reason for social weighting. If we are averse to inter-temporal inequality (as we show when we adjust for consumption growth with the ηc term in the Ramsey discount equation), consistency and ethics suggest that we act on an aversion to income inequality between countries as well as within them. Harm done to the poor should be weighted more heavily than harm done to the rich. Now, however, the shoe is now on the other foot. From the perspective of today, we are the poor generation compared to future generations, and discounting redresses that. But from the perspective of today's industrial countries, we are the rich and developing countries are the poor. Aversion to inequality implies that we weight prospective damages in developing countries more heavily than damages that we ourselves expect to experience. To weight consumption levels in aggregating over time, which is what η does in the Ramsey equation, but *not* to weight in aggregating across space, which social

GDP, whereas the richest 25 percent would *gain* up to 0.1 percent. These results were for projected global average temperature increases of 4°C and 5°C and an "experimental" approach to climate sensitivity estimates. If non-market effects such as lost lives and disruption of ecosystems had been included, the disparities would have been even greater. See Mendelsohn et al. (2006).

[24] Using existing purchasing power exchange rates. Differing capacity for adaptation between rich and poor will work toward maintaining inequality. Brekke and Johansson-Stenman (2008) agree that in the case of global warming, the distribution of income within generations should be considered, and provide some numerical examples.

weighting would do, has a whiff of hypocrisy. Not to put too fine a point on it, it is curious that inter-temporal weighting a la Ramsey is standard analysis and works to the advantage of our generation. Intra-generational weighting, which would work to the advantage of the future poor, is the exception.

In Practice

The mechanics of social weighting have been worked out for some time.[25] For the iso-elastic utility function we have been discussing, the distributional weight D attached to abatement costs or benefits incurred by country i can be given as

$$D_i = [C/C_i]^\eta,$$

where C is world average consumption level, C_i is the average consumption level in the country or region in question, and η is the (over-worked) inequality aversion parameter, (i.e., minus the elasticity of MU with respect to consumption). For example, if η is 2, C is \$10,000, and C_i is \$2,780, which are current world average and Indian per-capita GDP (ppp), D is 12.94. This means that a dollar of damages in country i (India in this illustration) is weighted at \$12.94. These weights will shrink if incomes converge. If India reaches world average income, the weight becomes 1. Social weights can also be normalized on OECD consumption levels and should lead to the same estimate of optimal emission levels (Azar 1999). Weights are time-dated to reflect changes in world income distribution, which is expected to converge. Anthoff, Hepburn, and Toll (2009) suggest that damage estimates be normalized on the MU of consumption in the country in which abatement measures are contemplated, to avoid confusion. If weights are normalized on world average per-capita income, as in the earlier illustration, it is possible that in BC analysis, the weighted damages would be (incorrectly) compared to unweighted abatement costs, as if they were in the same units. As explained later, however, there is a problem with this suggestion.

Complications

Despite strong reasons supporting social weighting, there are at least three reasons for caution. First, the choice of weights is, in large

---

[25] Squire and van der Tak (1975). See also Azar and Sterner (1996).

measure, arbitrary. At best, the weights will reflect our preferences, not necessarily the preferences of generations to come. Second, in principle, social weighting should be applied across the board to all government activities that are subject to BC analysis and that have substantial impact on poor countries, if it is applied anywhere. Otherwise one is in the awkward position of valuing, say, crop loss to a poor farmer from extreme heat differently than crop loss for that farmer due to plant disease that is unrelated to climate. Efficiency in allocating resources to prevent crop loss, or any loss, needs a standard metric.[26] This, however, may be impractical.

Third, if cost-benefit analysis is done and the benefits of abatement (damages avoided) are weighted, then abatement costs need to be weighted as well.[27] This introduces a serious problem. How are the costs of abatement to be allocated among countries? With country specific weights, different allocations of abatement costs will be weighted differently and will sum to different amounts. Every allocation will yield a different benefit/cost ratio, and the ability of BC analysis to guide policy will be compromised. It is possible that a particular abatement target could easily pass a BC test with one cost allocation scheme, but fail with another. One implication is that the greater the share of mitigation costs picked up by rich countries, the higher should be the optimal level of mitigation effort. Notice that the problem arises only on the cost side, where the allocation of abatement costs among countries is not settled, but is up for negotiation. In contrast, the distribution of benefits (damages avoided) is not negotiated, but is determined by the economic, climate, and physical features of countries.[28]

[26] This is also true for valuing deaths prevented. As explained in Chapter 2, standard practice is to estimate the value of a statistical life based on willingness and ability to pay. To avoid the odious result of weakening climate policy because deaths occur disproportionately in poor countries, social weighting can be used. This, however, implies social weighting of all efforts to reduce deaths in poor countries.

[27] This is a two-step process. Future damages at time t are weighted according to the income levels of those damaged and then discounted at the SRTP to calculate their present value. Mitigation costs are weighted according to the income levels of those taking mitigation actions at the time the measures are taken, and then discounted at the SRTP.

[28] The problem of cost allocation does not arise for most BC analyses, which are done at the national or subnational level. First, national governments have the authority and tools to shape income distribution, so social weighting may not be necessary. Second, the funding of projects and thus their cost allocation is under government control and not subject to negotiation among sovereign nations.

This third problem has a further twist. In BC analysis, least-cost methods are assumed to be used to obtain objectives. In the context of global warming, least-cost implies that the marginal cost of abating greenhouse gas emissions is equalized across countries. This can be accomplished by an internationally uniform carbon tax or a cap-and-trade system. Thus for efficiency, the distribution of abatement *effort* as among countries reflects their marginal abatement cost functions. But the allocation of *effort* can and should be separated from the allocation of *paying* for that effort. This is similar to the burden-sharing issue much discussed in the national-defense literature. Without social weighting, the least-cost distribution of abatement effort is independent of who finances the abatement. Once we introduce social weighting of abatement costs, this result becomes ambiguous. Specifically, different international financing arrangements for paying for abatement will lead to different global abatement cost functions, and different optimal levels of abatement. The greater the fraction of abatement costs borne by rich countries, the larger is the optimal level of abatement. Conversely and from an equity standpoint – and perversely – the larger the share on abatement costs shouldered by the poor, the weaker is the optimal abatement effort. The optimal level of abatement at the global level is tied to who is picking up the tab.

Anthoff, Hepburn, and Tol (2009) recognize the need to use the same metric for costs and benefits. Their solution, noted earlier, is that damage estimates, which have undergone social weighting, are to be normalized with the MU of consumption in the region in which the abatement project is to be undertaken. In this fashion, the costs and benefits are in the same metric. This works only if abatement activities are financed by the country in which they are pursued. However, as demonstrated in Chapter 8, international transfers of one type or another are essential for even a second-best post-Kyoto agreement (they are already taking place through the Clean Development Mechanism). This leaves us with the problem that different financing arrangements will lead to different optimal levels of abatement, and to multiple BC ratios.

The underlying problem is that the utility value of money differs internationally. The Anthoff, Hepburn and Tol (2009) study confirms that the MU of a dollar increases as it flows from global North to global South. Using a standard scenario and assuming the pure rate of time preference, ρ, to be 1 percent, they find that after weighting, the social cost of a ton of carbon is $47 when normalized for the United States, $3.40 for China, and $0.60 for Sub-Saharan Africa. This means that the present value of global damages from a ton of carbon emissions *when seen from the perspective of China* is about 7 percent of the value as seen from the United States.[29] No wonder reaching a post-Kyoto agreement is so difficult.

The problem for BC analysis is that until the allocation of abatement cost financing is fixed, the optimal level of abatement is indeterminate. One way to finesse this problem is to seek out Lindahl prices for allocating costs. As explained in Chapter 8, Lindahl prices are a device for allocating the costs of supplying a public good so that all countries are satisfied with the fixed supply of the good. Cost shares depend on willingness to pay, which in the case of global warming would take account of the benefits (damages avoided) a country enjoys, as well as its ability to pay. If these prices (cost shares) could be found, the pattern of abatement effort could be separated from the pattern of financing. The efficiency rule of equating marginal abatement costs could be preserved, social weighting would be unnecessary, and the problem of multiple BC ratios would be avoided. The remaining problem, examined in Chapter 8, would be to avoid free-riding and obtain universal participation in such an arrangement.

Conclusions

Once again, BC survives, but just barely. The overarching problems are how to keep discounting from trivializing the future, how to meld inter-generational efficiency and equity concerns, the tension between revealed preference data and ethics in determining discounting parameters, the inadequacy of the conventional utilitarian model to reflect multiple preferences concealed in η, the daunting range of plausible parameters, and finally the ambiguity of BC ratios when there is social weighting and no clear rule for allocating abatement costs. Solutions may include replacing the utilitarian model, breaking up η, turning to

---

[29] It does *not* mean that damages in China are 7 percent of U.S. damages, which is an entirely different question. See also Anthoff (2009).

declining discount rates, recalculating damages under environmental scarcity, and serious attention to using Lindahl prices.

References

Anthoff, D. (2009). Optimal Global Dynamic Carbon Taxation. *ESRI Working Paper* 278. February.

Anthoff, D., C. Hepburn, and R. S. J. Tol (2009). Equity Weighting and the Marginal Damage Costs of Climate Change. *Ecological Economics* 68 (3): 836–49.

Anthoff, D., R. S. J. Tol, and G. Yohe (2009a). Risk Aversion, Time Preference and the Social Cost of Carbon. *Environmental Research Letters* 4 (2): 1–7.

(2009b). *Discounting for Climate Change.* http://economics-ejournal.org/economics/journalarticles/209–24

Archer, D. and V. Brovkin (2008). The Millennial Atmospheric Lifetime of Atmospheric CO_2. *Climatic Change* 90: 283–97.

Atkinson, G., S. Dietz, J. Helgeson, C. Hepburn, and H. Saelen (2009). Siblings, Not Triplets: Social Preferences for Risk, Inequality and Time in Discounting Climate Change. *Economics: The Open-Access, Open-Assessment E-Journal* 3: 2009–26.

Azar, C. (1999). Weight Factors in Cost-Benefit Analysis of Climate Change. *Environmental and Resource Economics* 13: 249–68.

Azar, C. and T. Sterner (1996). Discounting and Distributional Considerations in the Context of Global Warming. *Ecological Economics* 19 (2): 169–84.

Brekke, K. A. and O. Johansson-Stenman (2008). The Behavioral Economics of Climate Change. *Oxford Review of Economic Policy* 24 (2): 280–97.

Boardman, A., D. Greenberg, A. Vining, and D. Weimer (2006). *Cost-Benefit Analysis: Concepts and Practice.* Upper Saddle River New Jersey: Pearson/Prentice Hall.

Cowell, F. A. and K. Gardiner (1999). Welfare Weights. *OFT Economic Research Paper.* Office of Fair Trade, London.

Dasgupta, P. (2001). *Human Well-being and the Natural Environment.* Oxford: Oxford University Press.

(2007). The Stern Review's Economics of Climate Change. *National Institute Economic Review* 199: 4–7.

(2008). Discounting Climate Change. *Journal of Risk and Uncertainty* 37 (2–3): 141–69.

Dietz, S. and N. Stern (2008). Why Economic Analysis Supports Strong Action on Climate Change. *Review of Environmental Economics and Policy* 2 (1): 94–113.

Evans, D. J. and H. Sezer (2005). Social Discount Rates for Members of the European Union. *Journal of Economic Studies* 32 (1): 47–59.

Fankhauser, S., R. S. J. Tol, and D. Pearce (1997). The Aggregation of Climate Change Damages: A Welfare-theoretic Approach. *Environmental and Resource Economics* 10 (3): 249–66.

—— (1998). Extensions and Alternatives to Climate Change Impact Valuation: On the Critique of IPCC Working Group III's Impact Estimates. *Environment and Development Economics* 3: 59–81.

Gollier, C. (2002). Discounting an Uncertain Future. *Journal of Public Economics* 85: 149–66.

Groom, B., C. Hepburn, P. Koundouri, and D. Pearce (2005). Declining Discount Rates: The Long and the Short of It. *Environmental and Resource Economics* 32: 445–93.

Guo, J., C. Hepburn, R. S. J. Tol, and D. Anthoff (2006). Discounting and the Social Cost of Carbon: A Closer Look at Uncertainty. *Environmental Science and Policy* 9: 205–16.

Ha-Duong, Minh and N. Treich (2004). Risk Aversion, Intergenerational Equity and Climate Change. *Environmental and Resource Economics* 28: 195–207.

Hepburn, C. (2006). Discounting Climate Change Damages: Working Notes for the Stern Review. *Mimeo*.

Hoel, M. and T. Sterner (2007). Discounting and Relative Prices: Assessing Future Environmental Damages. *Climatic Change* 84: 265–80.

Hope, C. (2008). Discount Rates, Equity Weights, and the Social Cost of Carbon. *Energy Economics* 30 (3):1011–19.

Koopmans, T. (1967). Objectives, Constraints and Outcomes in Optimal Growth Models. *Econometrica* 35 (1): 1–15.

Kverndokk, S. and A. Rose (2008). Equity and Justice in Global Warming Policy. *International Review of Environmental and Resource Economics* 2: 135–76.

Mendelsohn, R., A. Dinar, and L. Williams (2006). The Distributional Impact of Climate Change on Rich and Poor Countries. *Environment and Development Economics* 11 (2): 159–78.

Newell, R. and W. Pizer (2003). Discounting the Distant Future: How Much Do Uncertain Rates Increase Valuations? *Journal of Environmental Economics and Management* 46: 52–71.

Nordhaus, W. (2008). *A Question of Balance: Weighing the Options on Global Warming Policy*. New Haven, CT: Yale University Press. Prepublication version at http://www.nordhaus.econ.yale.edu/Balance_prepub.pdf

Pearce, D., B. Groom, C. Hepburn, and P. Koundouri (2003). Valuing the Future: Recent Advances in Social Discounting. *World Economics* 4(2): 121–41.

Pearce, D. and D. Ulph (1999). A Social Discount Rate for the United Kingdom. In *Economics and the Environment: Essays in Ecological Economics and Sustainable Development* (pp. 268–85), D. Pearce (ed). Cheltenham: Edward Elger.

Portney, P. and J. Weyant (eds.) (1999). *Discounting and Intergenerational Equity.* Washington, DC: Resources for the Future.

Ramsey, F. (1928). A Mathematical Theory of Saving. *Economic Journal* 38 (152): 543–59.

Saelen, H., G. Atkinson, S. Dietz, J. Helgeson, and C. Hepburn (2008). Risk, Inequality, and Time in the Welfare Economics of Climate Change – Is the Workhorse Model Underspecified? *University of Oxford Economics Department Discussion Paper 400.*

Sen, A. K. (1967). Isolation, Assurance, and the Social Rate of Discount. *Quarterly Journal of Economics* 81: 112–24.

Squire, L. and H. van der Tak (1975). *Economic Analysis of Projects.* Washington, DC: Johns Hopkins University Press for the World Bank.

Stern, N. (2007). *Stern Review: The Economics of Climate Change.* Cambridge: Cambridge University Press.

Sterner, T. and M. Persson (2008). An Even Sterner Review: Introducing Relative Prices into the Discounting Debate. *Review of Environmental Economics and Policy* 2 (1): 61–76.

Weitzman, M. (1994). On the "Environmental" Discount Rate. *Journal of Environmental and Economic Management* 26 (2): 2000–9.

(1998). Why the Far Distant Future Should be Discounted at the Lowest Possible Rate. *Journal of Environmental Economics and Management* 36: 201–8.

(2010). GHG Targets as Insurance Against Catastrophic Damages. *June, Mimeo.*

4

Empirical Estimates

A Tasting Menu

This chapter describes how numbers used in benefit cost (BC) analysis materialize and are assembled. They are not simply plucked from the air, but they are not as solid as we would like, either. Our purpose is not an exhaustive description, but rather to convey the flavor of the process. We start with a brief introduction to Integrated Assessment Models (IAMs), the main tool for empirical investigation of climate policy. At the center of IAMs are damage functions that identify the impacts of climate change and attempt to put monetary values on them. Damage functions are the spline that joins the science and economics of climate change. We then dig down one level to understand where the numbers in two specific impact areas, agriculture and sea-level rise, come from. The costs of global warming include adapting to higher temperatures as well as residual damages. We continue by reviewing some recent attempts to estimate adaptation costs. Finally we look into a specific proposed response to global warming – forestry policy – from the perspective of BC analysis.

Integrated Assessment Models

All climate IAMs have two basic features. They are designed to illuminate the interconnected physical, biological, and socioeconomic elements of climate change. They are also designed to assist in policy formation. IAMs come in two flavors: policy evaluation models (simulation models), and optimization models. The first assesses the physical, ecological, or economic effects of an exogenously proposed climate policy. The second seeks out an endogenously derived optimal policy to maximize welfare or minimize cost. Optimization models generally are less detailed and less complex than policy evaluation models so as not to exceed computational constraints. While simplification involves a cost, it assists in transparency and understanding.

At the core of a welfare optimizing IAM is an optimal economic growth model to which a carbon cycle model and a climate model have been attached. The starting point is scenarios for population growth and economic productivity, which determine economic output paths. Together with estimates of the energy composition and intensity of production, economic output determines emissions. Taking into account ocean uptake of carbon emissions and estimates of carbon released from land use changes, greenhouse gas emissions trajectories are estimated. Some IAMs are limited to CO_2, whereas others (e.g., FUND) include other greenhouse gases such as methane. The emissions are inputs into a carbon cycle or greenhouse gas model, which converts emission flows – which are in tons per year – to atmospheric concentrations measured in parts per million or parts per billion. This calculation takes into account the rate at which carbon is removed from the atmosphere, mostly through absorption by the oceans. The concentrations of carbon and other greenhouse gases are then linked to temperature changes in the climate model. IAMs do not contain full, complex climate models, but try to capture the essential links between concentrations and temperature in a reduced-form fashion. This step is the climate sensitivity issue and is subject to considerable scientific uncertainty.[1] It is also the point at which feedback from higher temperatures to additional emissions of carbon from melting permafrost and release of deep ocean methane hydrates might be introduced.

The next steps are to link the climate model back to the economic model by estimating the physical and ecological effects of warming and putting monetary values on them when possible. (Some effects may be beneficial, for example reduced heating costs in high latitudes. They can be viewed as negative costs.) We look at this valuation step more closely in the next section. The time-dated, monetized damages plus estimates of cost-effective adaptation measures can then be discounted and summed to obtain the present value of global-warming

---

[1] The climate sensitivity parameter measures the impact on temperature of a doubling of greenhouse gas concentrations. See further discussion in this chapter.

damages under a business-as-usual (BAU) assumption. Alternatively, one can calculate the social cost of carbon – the present value of the stream of damages arising from one additional ton of carbon released into the atmosphere.

The next task is optimization, and this requires an objective function – the inter-temporal maximization of welfare. This is most commonly a utilitarian social welfare function (see Chapter 3). The control variable to maximize utility is the optimal division of economic output over time to consumption, investment, and abatement of greenhouse gases. At this point, an abatement cost function is needed. The one selected usually assumes an efficient allocation of abatement effort as among countries, sectors, and greenhouse gases, but alternatives can be explored. Abatement cost functions are based on (uncertain) technology projections. The computer then reveals the optimum trajectory of greenhouse gas emissions so that at each point in time the marginal costs of abatement are just equal to the marginal benefits (marginal global-warming damages avoided). The time path of the marginal benefits is the same as the time path of the social cost of carbon and provides a guide for specific optimal policy measures (carbon taxes or cap-and-trade). Models are constructed so that it is also possible to calculate the BC ratio for various temperature or atmospheric concentration targets.

These are the bare bones of an IAM. There is plenty of scope for variation. One feature is whether adaptation policies are explicitly modeled. As discussed in the next chapter, mitigation and adaptation are (imperfect) substitutes. Another important difference is in the degree of regional disaggregation. Some construct regional climate-economy models that track regional income and consumption growth as well as regional damages. While more data-intensive, regional disaggregation allows for region-specific discount rates (which depend in part on regional rates of growth of consumption); social weighting based on income disparities among countries (regions); and region-specific damages and adaptation capabilities. The regional and national incidence of benefits and costs of abatement are an essential input into models that investigate international cooperation in international environmental agreements (see Chapter 8).

The third distinguishing feature is the treatment of technology. It is universally acknowledged that technological changes are critical to understanding how climate policies will unfold. This is especially problematic for modeling, as technology itself has attributes similar to public goods. IAMs typically assume an exogenous trend in production technology, energy intensity, and so on. Some go beyond this and include technology spending (R&D) as a policy option that, together with rising carbon prices, drives technological change. The impact of technology on the cost of alternatives to fossil fuels, the availability of backstop technologies, technological learning and dissemination, and the form and effectiveness of technology policy are all at issue.

Another feature that distinguishes IAMs are whether they are restricted to direct economic effects – for example, agriculture, tourism, and the like – or attempt to include indirect impacts of economic significance – human life, health, biodiversity, and so forth. There is an obvious trade-off here. Market effects are likely to be more solidly grounded, but some studies show that indirect effects can be very large, and to omit them could weaken the analysis. Along these same lines, individual IAMs offer greater or lesser detail on the specific sectors that are affected and on the energy sources and markets that give rise to emissions.

Early generations of IAMs were criticized for inadequate treatment of uncertainty. They were also criticized for brushing aside the possibilities to minimize total global-warming costs by cost-effective adaptation measures. Finally, they were criticized for either ignoring or superficially treating low probability–high damage scenarios – catastrophe modeling. The first shortcoming was thought to unduly minimize global-warming damages and the need for strong abatement measures. The second was thought to exaggerate the damages of global warming and thus exaggerate the need for strong abatement measures. Attention to the third alleged weakness, inadequate treatment of catastrophe, has tended to rehabilitate arguments for early and strong abatement measures. Recent and updated IAMs attempt to correct these limitations.[2]

Damage Functions: The Weakest Link?

At the heart of optimizing IAMs are three parameters: climate sensitivity (the change in global temperature resulting from a doubling of

---

[2] See, for example, the updated PAGE model and its treatment of adaptation opportunities and "large scale discontinuities" (Hope 2006).

atmospheric CO_2e above pre-industrial levels); the social rate of time preference, which is based on a pure rate of time preference and on an inequality-aversion preference; and a parameter relating temperature change to monetized damages. These are also the most important factors contributing uncertainty to estimates of the social cost of carbon.[3] Climate sensitivity is based on science and is investigated through global circulation models. The social rate of time preference was discussed earlier in Chapter 3. This leaves the relation between temperature and damages. Can reasonably accurate estimates of damages (the benefits of abatement) be obtained? The estimates need not be precise to be useful. However, policy made on the basis of highly inaccurate numbers would be perilous.

Pearce (2005) offers a good description of how damage functions are constructed in optimizing IAMs. The starting point is an estimate of the temperature increase that would result from a doubling of CO_2e concentrations. Call this parameter "A". Pearce uses a value of 2.5 °C in his example. The IPCC Fourth Assessment Report places it between 2 °C and 4.5 °C, with a probability exceeding 66 percent. The damages associated with this temperature increase are estimated by using bottom-up regional and sector studies, described later. The damages from a doubling of CO_2e at some time t in the future are the damages from doubling today's concentrations, scaled up by population growth and increasing willingness to pay because of higher per-capita incomes.[4] Call these future doubling damages, k_t. k_t is then multiplied by the ratio of temperature increase at t, to the parameter A, and the ratio raised to the power λ. In the original work, Fankhauser (1995) gave λ a range of 1 to 3. Thus if temperature at time t is expected to increase by 5 °C, and if the scaled-up damages, k_t, will be $100, and we select A = 2.5 and λ = 2, we would calculate the damages at time t to be $400.[5] These damages would then be discounted to get their present

---

[3] Jamet and Corfee-Morlot (2009, Table 6) attempt to rank them by importance.

[4] Pearce suggests an income elasticity of willingness to pay for avoiding damages of 0.3–0.4.

[5] Damages = $k_t[T_t/A]^\lambda$ where T is the estimated temperature increase at time t. This could be further adjusted to reflect the rapidity of temperature change – the more rapid, the more damaging. Stern (2007) uses a value of 1.3 for λ but argues in the technical appendix that 1.3 is probably too low. The influential DICE 2007 model links monetized damages and temperate increases through a quadratic function (Nordhaus 2008) A 3 °C increase is estimate to cause damages of about 2.5 percent of global output, and a 6 °C change to cause damages equal to about 10 percent of global output.

value. All of the parameters are potentially important. The higher the estimate of current damages from a doubling, the higher the future damages. The higher the growth rate of population and the larger the elasticity of willingness to pay for damage avoidance, the higher the future damages. And the larger the exponent λ, the higher the future damages.

The conventional approach is to exploit existing sector- and country-specific studies looking at the damage effects of temperature change arising from a doubling of CO_2e concentrations, and to extrapolate these finding to concentrations and temperature changes above and below the doubling benchmark. The extrapolation is problematic as we have no real world data on damage costs should temperature increase by $4\,^\circ C$ or $6\,^\circ C$. There is an even more fundamental question as to the form that the damage function should take. Weitzman (2010a, 2010b) argues that an additive form of the damages function is as reasonable as the multiplicative form now commonly used, and would show much more serious damages at high temperatures. In any event, it is helpful to dig below the damage function itself to see how impacts are quantified and monetized. Before doing so, however, it is useful to note what the various studies conclude regarding marginal damage costs (the social cost of carbon).

An Uncertain Bottom Line

As noted in Chapter 3, the social cost of carbon is the single most important variable for policy analysis. It is the net present value of the sum of the stream of damages arising from the release of one more ton of carbon. If calculated from an optimization model, it reflects the Pigouvian tax that would correct the carbon emissions externality. Some 232 estimates have been published. Tol (2008, 2009) has performed a very useful meta-analysis of these estimates. The results show that, as expected, studies that used a relatively high pure rate of time preference (3 percent) had far lower mean estimates of the social cost of carbon than did studies using a zero rate ($18 versus $232 per ton carbon, both using 1995 dollars). Perhaps more important is the uncertainty surrounding the estimates. Whereas the mean for the entire sample of 232 estimates is $105, the mode is $13, implying some very high estimates. Indeed, at the ninety-fifth percentile, the social cost of carbon jumps to $360 and at the ninety-ninth, to $1,500. Further statistical analysis tends to confirm that the probability distribution function has a fat right tail, exactly the issue that has concerned Weitzman and others and which tends to undermine the credibility of conventional cost-benefit analysis (see Chapter 2).[6]

Generating the Numbers

The Art of Shadow Pricing
In principle, valuation of environmental impacts is based on willingness to pay to avoid damages or willingness to accept compensation for incurring damages. There is an immediate problem. Many empirical studies show that willingness to pay for a small improvement in environmental quality is less than willingness to accept compensation for a small deterioration in quality. If we ask ourselves how much we are willing to pay to improve future climate over what it otherwise would be, the amount may be significantly smaller than if we were able to ask future generations what they would be willing to accept for its deterioration. However, future generations cannot be consulted, and a potentially important underestimation bias may be present.

For commercialized environmental goods and services, market prices are the first approximation of willingness to pay. Ecotourism is an example. If cleverly managed, nature can provide services that are priced and marketed to domestic and international tourists. For non-marketed goods and services, the traditional valuation tools use so-called surrogate markets or contingent valuation methods to estimate shadow prices. For example, even though there is no direct market for many environmental amenities, their value may be inferred from statistical relations between an amenities index, say smog or noise, and property values. The property market is the surrogate market. In contingent valuation, people are surveyed about their willingness to pay or accept compensation for some specific change in the environment,

---

[6] It would be incorrect to directly compare these estimates of the social cost of carbon with current prices in the EU trading market and infer that the EU is over- or under-pricing carbon. The studies generally assume an efficient response to global warming with a single worldwide carbon price. At the moment, only the EU is pricing carbon through its cap-and-trade scheme.

for example rehabilitating a degraded trout stream. These tools are not perfect, but there is extensive experience and considerable confidence in them.

Unfortunately, these tools cannot be directly applied to future damages from global warming. We do not know what market prices will be 50 or 100 years from now, and we cannot ask those not yet born what their preferences will be. In these circumstances, there are two options. One is to ask how the future impact would be valued if it were to occur today, and then project that value forward based on income growth estimates and conjectures about income elasticity of demand. For example, we have some data on the value of a statistical life in countries at different income levels. We might use this to calculate income elasticity and, together with per-capita income projections, estimate the value of lives lost to increased flooding or severe storms or heat stress at some time in the future. The second option is to make more explicit assumptions about what the future will look like. For example, it is not unreasonable to assume that through scientific progress and higher incomes, the incidence of tropical diseases will decline. In that case, it would be incorrect to ask what the damages from increased disease would be today. The impact itself – the incidence of temperature-related disease – will be different in the future.

There are more specific issues. First, most detailed damage studies have been made for developed countries. Whether the results can be transferred to developing countries where ecological, economic, and institutional factors differ is questionable. This is especially important in studies that emphasize adaptation, as there is strong evidence that adaptation capabilities are closely related to income levels. Second, most research follows an enumerative approach, adding up damages sector by sector. Adding up may miss collateral damages in interrelated sectors, but it can also double-count damages common to more that one sector. For example, rising sea levels contribute to saltwater intrusion in coastal aquifers and have negative effects on agriculture. Those damages may be overlooked in considering agricultural damages, but on the other hand could show up twice in the sea level and agricultural damage accounts. Perhaps more importantly, a general equilibrium rather than an enumerative approach is superior for capturing interactions between markets. For example, Bosello and his colleagues (2007) studied damages from sea-level rise in a modified computable equilibrium model. They found that including indirect effects significantly increased total damages and also led to quite different distributional effects. Third, not all studies take account of dynamic effects. Climate impacts and mitigation will likely impact long-term growth by affecting savings rates and capital accumulation. Fankhauser and Tol (2005) find that dynamic effects may be substantial as compared to the traditionally measured static effects, especially for developed countries. Fourth, in aggregating monetary values over countries and converting them to a common currency, a choice has to be made between market exchange rates and purchasing power parity exchange rates. We now drop down one level and look at specific sectors.[7]

Agriculture

Putting monetary values on the agricultural impact of climate change is facilitated by three factors. First, agriculture is marketed and prices are available so that gains and losses can be directly valued. Second, because agricultural yields are closely correlated to inputs (water, fertilizer, soil) and temperature, and because there exist wide regional variations in temperature, it is possible to infer a statistical relation between yields and temperature from existing data. Third, unlike some impacts, the effects of temperature can be subject to laboratory experimentation and analysis. At the same time, agriculture is subject to three complicating features. First, agricultural yields are very site-specific and the analysis should be done at a fine level of geo-spatial disaggregation. Second, up to some point, atmospheric carbon acts as a fertilizer promoting higher yields for some crops. Yield estimates respond positively to carbon fertilization and negatively to temperature increase, and it is difficult to disentangle them. Third, agriculture's share of output tends to be highest in very poor countries. This means the issue of social weighting takes on greater significance.

At this point we are interested in how damage (or benefit) numbers are generated. Cline (2007) explains and demonstrates two widely used approaches. The first is the Ricardian approach. It hypothesizes that land values capitalize the future stream of income generated by the land, and that the values are a function of temperature and

---

[7] Tol (2002a, 2002b) does a good job of assembling, standardizing, and interpreting damage estimates in different sectors. See also Tol (2009) for a review and assessment.

precipitation among other variables. Using data on agricultural output and climate variables from farm survey and county-level studies
across different climate zones, researchers have estimated statistical
models relating agricultural productivity to climate. Cline then uses
temperature and precipitation projections derived by averaging the
results of several general circulation climate models to calculate climate changes toward the end of this century. These changes are calculated by country and by regions within larger countries from a detailed
spatial grid. The two sets of data are brought together to estimate the
impact of global warming on agricultural output capacity, by country
and region. The actual calculations require multiplying the percentage change in rental value of agricultural land, as determined by the
models, times the ratio of net agricultural revenue (the return to land)
to the value of agricultural output, in order to obtain the percentage
change in agricultural output capacity.[8] Changes in output capacity are
also reported after a positive, if controversial, adjustment is made for a
carbon fertilization effect. The Ricardian approach relies on accurate
seasonal and spatial data on temperature. The variation can be substantial. For example, Hanemann (2008) notes that although a particular IPCC scenario anticipated a 2 °C global increase in temperature,
statewide California temperatures were expected to increase by 3.3 °C,
statewide California summer temperatures by 4.6 °C, and California
Central Valley temperatures by 5 °C. Whereas a 2 °C increase might be
expected to have a positive impact on output due to carbon fertilization, it is likely that a 5 °C increase would cause substantial losses.

Crop models constitute the second approach to monetizing damages. At their core they are estimates of production functions – output in physical terms as a function of an array of inputs including soil
types, nutrient levels, temperatures, water availability, solar radiation,
atmospheric carbon and other fertilizers, and other variables. The
models are developed in both laboratory conditions and in open field
trials. The U.S. Environmental Protection Agency (EPA) sponsored
a major crop models research effort under the direction of Cynthia
Rosenzweig and Ana Inglesia, and involving 125 agricultural research

---

[8] The ratios of net to gross agricultural revenue range from 0.41 for the United States
   to 0.78 for Africa, reflecting the higher input intensity of agriculture in developed
   countries.

stations worldwide. The principal crops were wheat, rice, maize, and soybeans, and the objective was to estimate the effects of global warming. The studies used a variety of global climate models and assumptions regarding adaptation responses. The results were the second main input into the Cline research effort.

The Cline study proceeds to use a weighted average of the Ricardian and crop model estimates to obtain what it calls "preferred" estimates of output gains and losses, with and without carbon fertilization, by country, in dollar terms and as a percent of each country's output. Toward the end of the century, the U.S. loses 5.9 percent of agricultural production without carbon fertilization but gains 8 percent with it. India loses 38.1 percent without carbon fertilization and loses 28.8 percent with fertilization. One takeaway conclusion is that the world as a whole will sustain a modest adverse impact on agriculture toward the end of this century under a BAU climate policy, and a potentially severe impact if carbon fertilization does not materialize and water scarcity limits irrigation. In 2003 dollars, the worldwide annual loss in the 2080s will be $187 billion without carbon fertilization and $38 billion with fertilization. The larger loss is about 16 percent of world agricultural production. A second conclusion is that agricultural losses will be heavily concentrated in Africa, Latin America, and India. One should remember that temperatures and losses will continue to rise in subsequent centuries.

Sea-Level Rise

Sea-level rise is an important source of global-warming damages. The 2007 IPCC Fourth Assessment Report projected a global mean sea-level rise of 18–59 cm over the balance of this century. This may be unduly optimistic. Rahmstorf et al. (2007) report that since 1990, sea levels have been rising faster than was projected by models: Satellite data show a linear trend of 3.3 mm per year plus or minus 0.4 mm for 1993–2006, whereas IPCC's 2001 projected best estimate was 2.2 mm per year. Because scientific understanding is limited, the 2007 IPCC report did not assess the likelihood of their estimate, nor provide an upper bound. Specifically they did not include the possibility of collapse on the West Antarctic and Greenland ice sheets, either of which could increase sea levels by 5-plus meters. Some research suggests that irreversible melting of the Greenland icecap could *ultimately* occur with temperature increases as low as 1.2 °C–3.9 °C (Garnaut 2008). Moreover, the IPCC estimate is a global average, and regional sea-level changes can deviate from the global mean by up to 100 percent (IPCC 2001). In any event, sea levels will continue to rise from ice melt and thermal expansion for the next 1,000 years even if CO_2 emissions peak and decline over the rest of this century.

The impacts and damages depend in part on the magnitude and speed of the rise, and on adaptation (protection) opportunities. The impacts include not only loss of lives, land, structures, and physical capital from inundation and storm surges, but also agricultural and fisheries losses, increased saltwater intrusion on surface and aquifer fresh water, and migration costs. At the same time, there may be substantial cost-effective adaptation measures protecting against sea-level rise. The cost of these measures should be considered part of the total damages. Unlike agriculture, there are no gains to be enjoyed.

We continue to be mainly interested in how damage numbers are constructed. We offer two examples, each adding a different insight. Susmita Dasgupta and her colleagues (2009) have set out to estimate the impacts of sea-level rise on eighty-four coastal developing countries. The end result is not a single damage number but a better understanding of the magnitude and international distribution of impacts that have economic significance. The impact categories are land lost to inundation, directly displaced population, GDP, urbanized area at risk, agricultural area at risk, and wetlands at risk. Impacts are calculated for sea-level increases from 1 to 5 m. No attempt is made to project future socioeconomic conditions, so the impact estimates are not dated. The effects of storm surge are not considered. The method is to construct from geophysical data the areas that would be inundated with sea-level increases of 1 to 5 m. The next step is to assemble data on the amount of land, population, GDP generated, urbanization, agricultural use, and wetlands area at risk from inundation, and overlay the impact maps on the inundation maps.

Both the global impacts and their distribution across countries are of interest. On a global basis, a 1-m sea-level rise will inundate 0.3 percent of the land areas of coastal countries, displace 1.3 percent of their populations, cause direct GDP losses of 1.3 percent, inundate 0.4 percent of agricultural land, and inundate 1.9 percent of wetlands. With a 5-m rise, the effects would be 1.2 percent loss of land area, 5.6 percent population displacement, 6 percent direct GDP loss, 2.1 percent loss of agricultural land, and 7.3 percent loss of wetlands. The impacts are highly concentrated in a small number of developing countries. To take an extreme example, even with the smallest sea-level rise – 1 m – 5.2 percent of Vietnam's land area would be under seawater, 10.8 percent of its population would be displaced, 10.2 percent of its GDP would be directly lost, and 7.1 percent of its agricultural land and 29 percent of its wetlands would be gone. These estimates, of course, fall short of actual damage estimates – adaptation measures, general equilibrium effects, and other factors are not calculated. At the same time, other global-warming effects such as decreased availability of fresh water and declining crop yields can amplify these effects. For example, Vietnam is expected to suffer 15 percent agricultural yield losses due to temperature increases, as well as decreased flow and greater seasonal variation in the Red and Mekong rivers.[9]

In our second example, Anthoff, Nicholls, and Tol (2010) use an IAM, FUND. Their analysis considers progressively more severe sea-level increases of 0.5, 1, and 2 m; adds a cost-benefit component; considers a variety of impact cost categories including displaced people and loss of wetlands; and breaks down the results by country or region. The BC analysis captures the optimal fraction of coastline to be protected. This balances the cost of protection against the costs of retreat. Protection costs include not only direct protection measures (building and maintaining dikes), but also the value of wetlands lost if protection is undertaken. The latter is the result of the so-called coastal squeeze that occurs when salt marshes and mudflats become trapped between seawalls and rising sea levels, and over time disappear. In essence, the sea walls prevent the wetlands from migrating landward. Wetland value is assumed a function of per-capita income in the country at the date when it is lost and based on an average OECD value of $5 million per km^2. The cost of a displaced person is assumed to be three times per-capita income in the country at the time of displacement. Annual protection costs depend on the rate of sea-level rise and the fraction of coastline protected. The benefits are dry land losses avoided.

[9] Cline 2007 (without carbon fertilization); Vietnam Ministry of Natural Resources and Environment (2003).

Several results are of interest: first, total damage costs, which include protection costs, are about five times as high for a 1-m sea-level rise as for a 0.5-m rise, and ten times as high for a 2-m rise as compared to a 0.5-m rise; second, of the four categories of costs – protection, dry land loss, wetland loss, displaced persons – protection costs dominate and wetlands loss is second; third, scenarios that project slow economic growth lead to higher levels of total damage costs (excluding wetland losses); fourth, total damage cost with no protection are about 3.5 times as high as with optimal protection for a 0.5-m sea-level rise, but this drops to about 1.4 times as high for a 2-m rise. The takeaway conclusions, subject to a number of caveats, are that widespread protection measures for developed coastal areas are more rational than is generally thought; that rapid economic growth encourages greater protection and reduces damages; and that some poor developing countries may have great difficulty in financing otherwise optimal protection measures.

Adaptation Costs

Adaptation costs enter economic analysis of global warming at three points. First, they are essential in determining the optimum amount of adaptation effort. This can be done on a very specific level – is a seawall to protect Hua Hin beach on the Gulf of Thailand justified? Should the Indonesian government stockpile tents in case of flood-related evacuees? These are typical, project-centered BC questions. But adaptation can also be folded into more aggregate studies. For example, both the Ricardian and crop model analysis discussed earlier accommodate some adaptation measures, as did the second sea-level rise example.

Second, adaptation costs are a component of total global-warming costs and are needed to determine the optimal amount of *mitigation* effort. As discussed in the following chapter, adaptation and mitigation are (imperfect) substitutes. It follows that the optimal levels of adaptation and mitigation in principle should be jointly determined. This does not need an unduly high degree of accuracy, but to completely neglect or misjudge adaptation opportunities and their costs can lead to big mistakes in mitigation policy.

Third, decisions on how international adaptation cost burdens are to be shared internationally help determine the willingness of countries to agree on global-warming limits (Chapter 8). For better or worse, negotiations for a post-Kyoto climate agreement have bound together the issue of emission reduction commitments and international financing to help with adaptation.[10] At the moment we are most concerned with how adaptation cost numbers are obtained, and will take up more subtle questions concerning mitigation and adaptation in the next chapter.

Adaptation costs are country, sector, and development level specific, and are difficult to aggregate. Moreover, they are difficult to distinguish from normal features of development. Examples are research on drought-resistant plants and on tropical disease. What fraction of expenditures on agriculture or on public health research should be attributed to the incremental effects of global warming? Also, many adaptation measures are undertaken in the private sector and are difficult to estimate. For example, climate-related shifts in cropping patterns are made by farmers and are outside public-sector budgets. Even more fundamentally, because the need for adaptation is contingent on the amount of mitigation undertaken, the magnitude of funds needed for adaptation is an ambiguous concept. With these caveats in mind, we offer a small sampling of the adaptation cost literature.

Agriculture: Nelson (2009) and other researchers at the International Food Policy Research Institute (IFPRI) attempt to estimate the additional expenditures needed to hold child malnutrition numbers in 2050 to the level they would reach if there were no global warming. The basic method is to use biophysical crop models centered on five crops – rice, maize, wheat, soybeans, and groundnuts – and a global trade model together with two climate scenarios (one wetter, one drier) to estimate the effects of global warming on yields and caloric consumption. This is converted to additional incidence of child malnutrition. The assumption is that yield declines and additional child malnutrition can be offset with additional investments in agricultural

---

[10] International institutions have anticipated the role they may play in mediating adaptation funds. Efforts to understand and estimate aggregate adaptation costs have been made by the UNFCCC, the World Bank, the OECD, and IFPRI. The domestic incidence of adaptation costs is also important for political economy analysis.

research, irrigation, and rural roads. Although the details are not given, there is a large body of evidence that links research, irrigation, and rural transportation to yield increases. The principal conclusion is that incremental expenditures in developing countries of about $7 billion annually could offset the child nutritional effects of climate change in 2050. Forty-three percent would come from irrigation investments, 38 percent from rural roads, and 18 percent from research and development. Note that this is not cost-benefit analysis as there is no attempt to match the expenditures with benefits. Note also that the $7 billion is linked to quasi-public goods – R&D, irrigation, roads – and private adaptation costs are not explicitly addressed. Note also that by climate-change standards, the time horizon – 2050 – is quite short.

Health: The UNFCCC (2007) has estimated a limited number of adaptation costs in the area of human health, focusing on malaria, diarrheal disease, and malnutrition in developing countries. The study concentrates on measures to be undertaken by public health systems but acknowledges that improvements in infrastructure and disaster management could also have important health benefits. The time frame is 2030, and two emission scenarios are considered, one leading to eventual CO_2e concentrations of 550 ppm and one at 750 ppm. The basic approach is to use a major 2004 World Health Organization (WHO) study that projects incremental cases of these three health effects due to global warming. The global annual incremental numbers for 2030 are 132 million, 4.6 million, and 22 million for diarrheal disease, malnutrition (stunting and wasting only), and malaria, respectively. These represent increases ranging from 2.5 to 10 percent above current levels for the three health effects. The estimated number of additional cases is then multiplied by averaged incremental prevention costs, taken from various World Bank studies. The results show annual adaptation costs in the range of $4–5 billion. They are likely to be underestimated due to other health impacts not included and for which adaptation or prevention is feasible, and for questionably low costs per case prevented. Similar to the IFPRI agriculture study, the benefits of the adaptation measures are not addressed. Adaptation through preventative measures will be only partially successful, and the costs of remaining illness and loss of life will be part of the residual costs of global warming but are not part of this exercise.

Infrastructure: UNFCCC also estimated a range of adaptation costs for protecting physical infrastructure investment that is expected to be made in the year 2030. The estimates involved three steps. First, global fixed capital formation in 2030 was estimated to be $22.3 trillion, based on economic growth projections. Second, the fraction of infrastructure investment that would be vulnerable to climate-change damages is estimated from two databases on annual historical losses, the Munich Re and the Association of British Insurance data, calculated as a fraction of the (then) current fixed capital formation. The Munich Re database covers only "large" weather catastrophes. The two vulnerability ratios are 0.7 and 2.9 percent, respectively. Third, the study borrows an estimate from the World Bank that the costs of upgrading and adapting infrastructure investment to negate the damages from climate change could add 5%–20% to investment costs. The three sets of numbers – projected gross fixed capital formation, the fraction vulnerable to climate change, and percentage increase in investment costs needed for adaptation – were multiplied together, giving an estimated annual adaptation cost in 2030 ranging from $8 billion to $130 billion. About two-thirds of the adaptation costs would be incurred in OECD countries because they have the preponderance of infrastructure assets.

This is a rather narrow way to estimate adaptation costs. Housing appears to have been excluded from the concept of infrastructure. Thus a very major source of damages, and large and potentially cost-effective opportunities for preventative adaptation, are not considered. This omission *might* be appropriate if the narrow purpose of adaptation estimates is to fix the amount of funds needed from international donors, as housing is generally privately financed by individuals. But if the purpose is to inform decisions about the optimal level of mitigation in a cost-benefit framework, failure to consider adaptation measures for housing will produce incorrect estimates of total damages, which are adaptation costs plus residual damages. Next, the focus in the study on physical infrastructure skirts over the costs of establishing and maintaining "soft" infrastructure for dealing with extreme weather impacts. These include communications, early-warning systems, emergency response training and facilities, help in restoration of essential services, and so forth. The study did not include the real costs (and benefits) of land use planning decisions in anticipation of global warming.

Finally, the UNFCCC does not consider the so-called adaptation deficit. This concept is particularly relevant for health and infrastructure adaptation estimates. The basic idea is that currently, before climate change exerts its pernicious effects, there is a very large gap between existing conditions in poor countries and conditions that might be considered satisfactory. In the context of infrastructure, it suggests that to respond rationally to the incremental threats of global warming, it is necessary to do something about the major deficiencies in *existing* infrastructure, as well as anticipate climate change in modifying new investment. To be specific, there is currently a very large amount of physical capital, including houses, that is situated in locations that are highly vulnerable to natural disasters –flooding, mudslides, typhoons, and the like. Cost-effective adaptation measures (preventative and remedial) should address protection of existing infrastructure as well as new capital formation. While one can sympathize with the desire to capture the *incremental* costs of protecting against climate change in new investments, adaptation costs for the existing stock may be of equal or greater importance. Once again, to get reasonably accurate estimates of the total damages from global warming – adaptation costs and residual damages – it would be desirable to consider adaptation costs for the existing stock of infrastructure as well as for new investments.[11]

In all, the UNFCCC study examined adaptation opportunities and costs in six sectors. It reported on only five. It was unable to find the methodology and sufficient data in the area of natural ecosystems to hazard an educated guess as to additional resources need for adaptation. For the five reported sectors, the total costs range from $49–171 billion in 2030, with $27–66 billion falling on developing countries. Researchers at the IIED (Parry et al. 2009) have provided a careful assessment of this study. Their overall conclusion is that the UNFCCC study may have underestimated adaptation costs by a factor of two to three. Recall also that the most serious effects of global warming

---

[11] IIED (2009) has estimated that the annual cost of removing the housing and infrastructure deficit in poor countries may be about $315 billion per year for twenty years, and the additional cost of upgrading that stock to meet climate change may cost $16–66 billion per year.

are not expected until the end of this century and beyond. In some respects, costs in 2030 might be considered a down payment on larger costs to follow.

Counting (on) Trees: Slowing Deforestation

We round out this discussion of how BC numbers materialize by considering the role of forests in moderating global warming. The numbers are potentially very encouraging. No new or exotic technology needs to be developed.[12] In principle, benefits could start almost immediate. Money would presumably flow in an efficient and equitable fashion from rich countries to poor countries where most deforestation is occurring. The collateral benefits of reducing deforestation – biodiversity and watershed protection, reduction of nutrient leaching and of downstream siltation and flooding, maintaining sustainable forest uses – may be large. But the legal, political, and institutional obstacles are also large and complex.

The basic idea is simple enough. Deforestation in the tropics is estimated to cause about 25 percent of anthropogenic carbon emissions and up to 17 percent of all greenhouse gas emissions. If this could be substantially reduced at reasonable cost, the prospect for controlling climate change would be greatly improved. This has led to numerous proposals wherein developing countries would "earn" carbon emission reduction credits for reducing deforestation and carbon emissions below some baseline level. These credits could then be sold to rich countries to count against their emission reduction obligations. The arrangement would then be similar to a sector-level Clean Development Mechanism. The general term for these schemes is Reduction of Emissions from Deforestation and [Forest] Degradation (REDD). It was expanded at Copenhagen to include accounting for conservation and sustainable management of forests, and is now known as REDD+. The term has an awkward ring and might better be called forest conservation. As in the earlier adaptation discussion, we are interested here in how the numbers are generated and whether forest conservation schemes can pass a BC test. Before turning to the numbers, however, it is useful to reflect on how forest conservation differs

---

[12] Monitoring technology, however, may be an exception.

from the standard prescriptions of a cap-and-trade or a carbon-tax scheme.

One difference is that it is a *subsidy* for not doing something (deforestation), whereas cap-and-trade or a carbon tax extracts a *penalty* for doing something (emitting carbon). In principle, forest conservation payments create and endorse property rights for the owners of carbon sequestered in forests (who may be private-sector or government entities). Cap-and-trade schemes create and endorse property rights to the atmosphere by governments, and require payments from those who would emit carbon. Again in principle, money flows in different directions in a subsidy versus tax system. This distinction is sometimes obscured in discussions of international cap-and-trade or tax schemes, as most of them also contemplate some mechanism to subsidize participation by poor countries (see Chapter 8).[13]

A second difference is that forest conservation is one of several mitigation strategies. At this level, BC analysis is seeking the least-cost methods of meeting an overall emission reduction target. It follows that an appropriate metric to evaluate the benefits of forest conservation is the concept of "costs avoided." For example, if the alternative to forest conservation is reducing fossil fuel emissions at a cost of $20 per ton of carbon, then the marginal benefit of a ton of emissions avoided through forest conservation is $20. Although this $20 must be compared to the opportunity cost of maintaining that ton sequestered in the forest, it avoids the difficult task of identifying and monetizing the damages this ton would have created had it been emitted.[14]

A third difference involves timing and permanence. All three approaches – carbon taxes, cap and trade, and forest conservation – involve incurring economic costs now in the expectation of benefits in the future, sometimes the far future. But trees not felled today can be felled tomorrow. Subsidy payments for forest conservation made today are hostage to a change of mind and behavior tomorrow by

[13] Economic theory is skeptical about using subsidies to induce firms to reduce pollution. One reason is that unlike pollution taxes, subsidies draw resources into an industry and, while an individual firm may reduce its pollution, total industry pollution may rise, not fall. This does not seem to be relevant to forest conservation subsidies. If new firms are drawn into the sector and plant trees, so much the better – each tree is a new carbon sink.

[14] More correctly, that task is undertaken elsewhere, when aggregate costs and benefits are estimated. See earlier sections of this chapter.

the owners of the sequestered carbon. If payment precedes performance, there has to be some assurance that the carbon locked up in trees will remain locked up for a very long time. A binding conservation covenant attached to the forested land might be feasible in a very few countries with impeccable trustworthiness credentials, but is unlikely in most. Alternatively, the payment could be annualized and be viewed as "renting" the sequestering services year by year, but then the long-term commitment of the buyer is in question. By and large, this problem does not arise in an international cap-and-trade or tax system as monitoring can tie payments directly with performance in the same time frame.[15]

The first step in putting numbers on a forest conservation scheme is to establish a deforestation baseline projection. Deforestation is not generally a wanton act of destruction but a rational attempt to maximize returns from land. The fact that land is subject to pervasive externalities, arising in large part from inadequate property rights, merely complicates the analysis. The basic premise is that land is shifted out of forests to alternative uses, mainly agriculture, in response to economic signals – demand for agricultural output versus timber and industrial wood output. The underlying determinants are population growth, technology change, input and output market prices, and so on, and environmental factors such as topography, soils, climate, and others that favor forests over agriculture. These determinants are the inputs into land use models. The same models may be used to estimate the shift of land out of agriculture back to forest use through reforestation. With baseline deforestation projections established by country and by smaller spatial units, the next step is to introduce various carbon prices and simulate deforestation rates with positive carbon prices. The expectation is the higher the price of carbon, the higher the level of forest conservation. Moreover, if markets are working well, the land conserved in forests will have the lowest opportunity cost in terms of agricultural use forgone. The reduction in hectares deforested

---

[15] Imagine an international system in which poor countries are compensated for reducing emissions from burning fossil fuels below a BAU baseline. If the reduction is 1 ton in year zero, the country would have no incentive to *increase* its emissions by 1 ton in the following year as it would lose any opportunity to earn additional payments. The ton of carbon is permanently removed from the atmosphere. Forest conservation is provisional.

(i.e., the hectares of forests conserved) is then converted to reduced carbon emissions using site-specific data on the sequestered carbon content of different forests. In this fashion, country- or region-specific marginal costs of carbon storage can be estimated. *Ceteris paribus*, if the marginal cost of carbon that remains sequestered in forests (its opportunity cost) is less that the marginal cost of other mitigation efforts, and if one has some assurance that the carbon sequestered is quasi-permanent, the forest conservation measures pass the BC test. In a perfect market, and if all countries faced a common carbon price, marginal carbon storage costs would be equalized and the forest conservation effort would be globally efficient.

Kindermann and his colleagues (2008) provide a good example of this type of analysis. They use three different forestry-agricultural models to estimate detailed baseline deforestation rates to 2030. Annual tropical deforestation ranges from 10.6 to 12.2 million ha, depending on the model. Carbon content per ha ranges from 48 to 132 tons per ha, depending on region and model. Baseline annual CO_2 emissions from deforestation in the absence of new policies were estimated to be 3.2 to 4.7 gigatons of CO_2 (a gigaton is a billion tons). For comparison, the global CO_2 emissions from all sources were estimated to be 28.4 gigatons in 2007.

CO_2 prices ranging from zero to $100 were then introduced to trace out marginal cost curves by model and by country. The results can be expressed in different ways. With $20 per ton, the models estimate that emissions would fall by 1.6–4.3 gigatons of CO_2 annually in the period between 2005 and 2030. Alternatively, a CO_2 price of $10 per ton would convert to a annual *rental* value of $85–252 per ha per year, which, the authors state, is quite often higher than agricultural rents in the same regions. Finally, the study compares the cost per ton of reducing emissions through forest conservation with the cost of alternative reduction measures focused on energy. At an assumed avoided cost of $9 per ton of CO_2 – the cost of the alternative methods to achieve a 550 ppm concentration target – forest conservation could reduce deforestation by 10–50 percent in the period up to 2030, making a substantial contribution to overall emission reductions in a cost-effective manner.

In certain respects, these should be considered idealized best-case results. The study assumes zero transactions cost, efficient implementation, and no carbon leakage (shifting deforestation from participating to non-participating regions). It assumes the emission reductions are permanent. On the other hand, it does not add in the multiple collateral benefits to the host countries and to the international community from forest conservation that are additional to carbon storage.

Conclusion

There is no question that many of the numbers discussed in this chapter are shaky. That does not make them worthless. It simply argues for caution in their use. And it invites critics to come up with better ones.

References

Anthoff, D., R. Nicholls, and R. S. J. Tol (2010). The Economic Impact of Sea-Level Rise. *Mitigation and Adaptation Strategies for Global Change* 15 (40): 321–35.

Bosello, F., R. Roson, and R. S. J. Tol (2007). Economy-wide Estimates of the Implications of Climate Change Sea Level Rise. *Environmental and Resource Economics* 37: 549–71.

Cline, W. (2007). *Global Warming and Agriculture: Impact Estimates by Country.* Washington, DC: Center for Global Development and Peterson Institute for International Economics.

Dasgupta, S., B. Laplante, C. Meisner, D. Wheeler, and J. Yan (2009). The Impact of Sea Level Rise on Developing Countries: A Comparative Analysis. *Climatic Change* 93: 379–88.

Fankhauser, S. (1995). *Valuing Climate Change: The Economics of the Greenhouse.* London: Earthscan.

Fankhauser, S. and R. S. J. Tol (2005). On Climate Change and Economic Growth. *Resource and Energy Economics* 27: 1–17.

Garnaut, R. (2008). *The Garnaut Climate Change Review.* Cambridge: Cambridge University Press.

Hanemann, W. (2008). What Is the Economic Cost of Climate Change? *Department of Agriculture & Resource Economics, University of California Berkeley, CUDARE WP* 1071.

Hope, C. (2006). The Marginal Impact of CO_2 from PAGE 2002: An Integrated Assessment Model Incorporating IPCC's Five Reasons for Concern. *The Integrated Assessment Journal* 6 (1): 19–56.

IPCC (2001). *Working Group 1 Report: The Scientific Basis.* Cambridge: Cambridge University Press.

Jamet, S. and J. Corfee-Morlot (2009). Assessing the Impacts of Climate Change: A Literature Review. *OECD Economics Department Working Papers No.* 691.

Kindermann, G. et al. (2008). Global Cost Estimates of Reducing Carbon Emissions through Avoided Deforestation. *Proceedings of the National Academy of Sciences* 105 (30): 10302–7.

Nelson, G. et al. (2009). *Climate Change: Impact on Agriculture and the Costs of Agricultural Adaptation*. Washington, DC: International Food Policy Research Institute.

Nordhaus, W. (2008). *A Question of Balance:Weighing the Options on Greenhouse Warming Policy*. New Haven, CT: Yale University Press. Prepublication version at http://www.nordhaus.econ.yale.edu/Balance_prepub.pdf

Parry, M. et al. (2009). *Assessing the Costs of Adaptation to Climate Change: A Review of the UNFCCC and Other Recent Efforts*. London: International Institute for Environment and Development and Grantham Institute for Climate Change.

Pearce, D. (2005). The Social Cost of Carbon and Its Policy Implications. *Oxford Review of Economic Policy* 19 (3): 362–84.

Rahmstorf, S. et al. (2007). Recent Climate Observations Compared to Projections. *Science* 316 (5825): 709.

Tol, R. S. J. (1999). The Marginal Costs of Greenhouse Gas Emissions. *Energy Journal* 1: 61–81.

(2002a). Estimates of the Damage Costs of Climate Change – Part 1: Benchmark Estimates. *Environmental and Resource Economics* 21 (1): 47–73.

(2002b). Estimates of the Damage Costs of Climate Change – Part 2: Dynamic Estimates. *Environmental and Resource Economics* 21 (2): 135–60.

(2008). The Social Cost of Carbon: Trends, Outliers, and Catastrophes. *Economics, the Open-Access, Open-Assessment E-Journal* 2 (25): 1–24.

(2009). The Economic Effects of Climate Change. *Journal of Economic Perspectives* 23 (2): 29–51.

United Nations Framework Convention on Climate Change (2007). *Investment and Financial Flows to Address Climate Change*. Bonn: UNFCCC.

Vietnam Ministry of Natural Resources and Environment (2003). *Vietnam's Initial National Communication Under UNFCCC*. Hanoi: Ministry of Natural Resources and the Environment.

Weitzman, M. (2010a). What Is the "Damages Function" for Global Warming – and What Difference Might It Make? *Climate Change Economics* 1 (1): 56–79.

(2010b). GHG Targets as Insurance against Catastrophic Damages. *NBER Working Paper 16136*.

5

Strategic Responses

This chapter marks a transition from our initial concern with the role of benefit cost (BC) analysis of mitigation to a broader focus on policy responses to climate change. The portfolio of responses starts with accelerated economic development and includes adaptation, a closer look at supply policy, technology, and extreme technology in the guise of geo-engineering. Each can add a useful element to the set of responses but mitigation (abatement) remains the center piece.

The Development Option

An argument can be made that the single most cost-effective response to global warming is accelerated economic development concentrated in poor countries. As discussed in Chapter 3, this strategy has a basis in efficiency and in ethics. Many (not all) economic studies suggest that in the near term, the return to investment in mitigating global warming is trumped by returns to more conventional development investments – physical infrastructure, health, education, and focused research and development. Investments that build productive capacity in developing countries also meet ethical concerns. They help alleviate poverty in the near term and help compensate future generations for climate damages by leaving them a larger productive base. Perhaps most important, all of the adaptation literature demonstrates that the capacity to adapt to higher temperatures is far greater in rich countries than in poor developing countries.

Timing may support a strategy of early efforts at rapid economic development to be followed by stronger mitigation efforts when the technology for reducing emissions has progressed. Nordhaus (2008) is clear on this. In choosing a discount rate that averages 4 percent over the next century (from 2006 to 2106), he calibrates his choice to data on the market return to capital and states that investments in controlling global warming should compete with investments in better seeds, improved equipment, and other high-yielding investments. His strategy does *not* imply that global warming is to be completely neglected. Indeed, in his optimal scenario, the social cost of carbon rises from $28 per ton today to $95 by 2050 and $202 by 2100. Nevertheless, the gist of the accelerated development argument is that by focusing initially on economic growth and development, mitigation technology is given time to improve *and* adaptation becomes more effective and carries a lower price tag.

One must be careful with this argument, however. First, technological progress in mitigation rests on a strong price signal in the near term. Second, there is no secure international mechanism through which resources not devoted to countering global warming will necessarily be made available for development, and no assurance that, if made available, they will be effective. Third, although poor countries may suffer disproportionately from global warming, rich countries have larger assets at stake. For them, accelerated development (growth) may provide more resources for subsequent adaptation, but the kernel of their problem is not underdevelopment and inadequate adaptive capacity. Their challenge is striking the right balance between substantial early expenditure on reducing emissions and adaptation costs in the future.

More broadly, it would be a mistake to attribute global-warming damages primarily to poverty. Poverty erodes the capacity to adapt to the effects of global warming but does not cause damages.[1] On the contrary, and everything else being equal, the more rapid economic development the higher the level of emissions, the greater the temperature increase, the greater the assets at risk, the higher the level of damages, and ultimately the more expensive and difficult will be adaptation. Global warming is not something that we can simply grow out of.[2]

---

[1] Deforestation associated with intensification of slash-and-burn agriculture may be an exception.

[2] IPCC scenarios show that the higher the economic growth rate, the higher global temperatures.

Economic growth without an explicit climate policy accelerates emissions, temperature increases, and damages. While the capacity to adapt is likely to improve, the externality problem that is at the center of global warming is left unaddressed.

Adaptation versus Mitigation

Three initial comments are relevant.[3] First, adaptation works best when uncertainty is least and climate change is slowest. Full advance knowledge of very slowly increasing temperatures would take much of the sting out of global warming. Second, mitigation and adaptation operate very differently. Mitigation changes the likelihood or intensity of an adverse event or process; adaptation changes the impact or consequences. Third, adaptation is the default strategy when mitigation fails to prevent temperature increase. Mitigation is not a default strategy.

Interest in adaptation has increased in recent years. This can be attributed to the growing realization that some global warming is inevitable even with heroic mitigation efforts. The prospect of international money for adaptation may also play a role. Why would it be easier to raise money for adaptation than mitigation? One reason might be that adaptation measures can "piggyback" on well-established structures for promoting sustainable development. Moreover, adaptation side-steps the hot-button issue of trade-competitive losses associated with mitigation via carbon restrictions. Third, the payoff for adaptation accrues to those who undertake it, not the international community. Small vulnerable countries may do better by concentrating on lobbying for adaptation funds rather than mitigation funds. In any event, this preference is still speculation – neither adaptation nor mitigation has raised large sums to date.

Adaptation to climate change adds a new, or newly fashionable, tool to the portfolio of climate policy responses.[4] Chapter 4 reviewed

---

[3] Stehr and Storch (2005) make the interesting observation that mitigation activities protect nature against society, and adaptation activities protect society against nature.

[4] For an early analysis, see Tol et al. (1998). By and large, IAMs have assumed that optimal adaptation is implicit in damage functions. This is unfortunate because it conceals the important question of the appropriate mix of mitigation and adaptation investments. One reason for the neglect is the lack of good aggregate data on

adaptation cost estimates. At this point, we wish to explore the extent to which mitigation and adaptation are substitutes and should be thought of as such in a cost-benefit framework.

At a high level of abstraction, they are, of course, substitutes. Consider the simplest case in which the only policy response to global warming is mitigation – reducing emissions of greenhouse gases. The optimal policy is then to pursue mitigation to the point where marginal mitigation costs are equal to marginal damages avoided (all properly discounted), and in which marginal mitigation costs are equalized as among countries, sectors, and greenhouse gases (converted to CO_2 equivalents). Mitigation occurs at the start of the chain from emissions to concentrations to temperature to climate to impacts to damages. Now introduce the possibility of adaptation – changes in socioeconomic conditions and behavior in anticipation or response to climate change. Adaptation occurs at the tail end of this chain, but the ultimate purpose is the same: to reduce damage costs. The task is to find the correct balance between the two.

The BC framework becomes slightly more complicated. Adaptation should be pursued up to the point where marginal adaptation costs are just equal to marginal damages avoided. At the same time, for efficiency, the marginal costs of reducing damages by one dollar from adaptation measures should be made equal to the marginal cost of reducing damages by one dollar by mitigation. In general, higher levels of mitigation reduce the amount of adaptation needed, and higher levels of adaptation reduce the need for mitigation. In this sense, they are substitutes.[5] In principle, mitigation decisions should be informed by adaptation opportunities and costs, and adaptation plans by mitigation plans. Ideally they should be jointly formulated. It may be helpful in thinking about an optimal program (as a benchmark, not necessarily practical) as minimizing total global-warming costs, consisting of mitigation costs, adaptation costs, and residual damage costs after

adaptation. Also, advocates of strong mitigation measures may sense defeat in considering adaptation.

[5] This is a somewhat facile statement and is only valid if the monetary values of both are accurate and commensurate. For example, mitigation reduces the risk of damages; adaptation acts less on risk and more on the magnitude of impacts. Mitigation reduces the need for internal migration; adaptation is making the best of misfortune. Apples and oranges can be substitutes if you are making juice but not if you are baking pies.

adaptation expenditures have been made.[6] We should also note that adaptation and mitigation will compete for funding, and an increase in resources to one may be at the expense of the other. This is the almost certain result of the Copenhagen Accord, which envisions a "Green Climate Fund" for mitigation and adaptation activities (see Chapter 9). Finally we note that in modeling optimal policy, there is a feedback mechanism from mitigation and adaptation expenditures, and in general, higher expenditures will slow economic growth, emissions, and subsequent damages.

For a full analysis, it may also be useful to distinguish two types of adaptation measures: those taken in anticipation of climate impacts and those taken in response to environmental impacts. Installing early-warning systems for severe storms is an example of the first, and rapid repair of utilities after flooding illustrates the second. One reason for the distinction is the timing. There can be substantial time lags between anticipatory expenditures (e.g., building sea walls) and avoided damages. Reactive expenditures are made at the time of impact. If this distinction is made, the proactive and reactive measures may be substitutes, in which case the marginal cost per dollar of damages avoided should be equalized as between the two types of adaptation. They may, however, act as complements in reducing damages. Early-warning systems *and* evacuation assistance may work best. Finally, the concept of anticipatory or proactive adaptation shades into the notion of adaptive capacity, and is discussed later.

The notion of substitutability between mitigation and adaptation should not be pushed too far. They are *imperfect substitutes* for a number of reasons. First decisions concerning mitigation and adaptation are made by different entities and at different levels. National governments decide emissions strategy and targets in international negotiations (e.g., Kyoto, Copenhagen). Functional ministries, local governments, and the private sector make adaptation decisions. Our earlier conclusion, that the decisions should be jointly formulated, is in large measure wishful thinking unless there is perfect two-way communication between the farmer in the field deciding what crops to

---

[6] de Bruin et al. (2009) and Tulkens and van Steenberghe (2009) use this framework, although the latter prefer the term "suffered damage costs" to the term residual damage costs."

plant and when and that country's delegation to Copenhagen. And for small poor countries, almost the only game in which they can now play is adaptation.[7] They have virtually no control over global emissions and hence mitigation. This would change dramatically with a global cap-and-trade system, in which they could sell emission reduction credits. Both mitigation and adaptation decisions would then at least be on the national agenda for these countries. But even then they should not be thought of as perfect substitutes. Mitigation decisions would be based on domestic opportunity costs and the international price they might receive for emissions reductions. Adaptation expenditures would be based on quite separate analysis of the costs and benefits of adaptation measures. The results of one set of calculations would be independent of the other.

Second, mitigation contributes to a global public good; by and large, adaptation benefits are secured at the national or local level. This makes a huge difference. As outlined in Chapter 8, the combination of international externalities and free-rider incentives makes funding mitigation extremely difficult. These obstacles are sharply reduced for adaptation, which tend to benefit the country, business, or individual who makes the investment. Even when there are public-goods aspect involved, as, for example, flood control or public health, national governments have the authority to tax and provide the public good. Whether there are sufficient resources to do so in poor countries is an important but separate question. Generally speaking, the absence of a global public good dimension for adaptation tilts the expenditure balance away from mitigation toward adaptation, even if mitigation were more cost-effective in reducing global damages. Governments prefer to spend for projects that benefit their own citizens. Also, because adaptation can be selective in protecting against specific impacts, it may be easier to mobilize funds for public works. Individuals, of course, will bear the brunt of adaptation costs. To put it a bit differently, unless the obstacles to a global mitigation effort are overcome, adaptation becomes the default policy.

Third, mitigation and adaptation differ in terms of timing. Mitigation costs are incurred in the near and medium term and avoid damages decades and centuries later. In contrast, reactive adaptation costs are

---

[7] Depending on the venue, they may, however, obstruct mitigation negotiations.

incurred at the time of impact. Anticipatory adaptation expenditures are likely to be relatively close in time to the benefits they provide. The economic implication of the difference in timing is that models that use high discount rates will favor adaptation, and models that use low discount rates will favor mitigation. In political economy terms, it may be easier to sell adaptation than mitigation because the time lag between cost and benefits is shorter and because it may be rational to defer adaptation spending, which is always easier on policy makers. Moreover, because of timing differences, mitigation does *not* become the default policy if adaptation is inadequate; mitigation and adaptation are asymmetrical substitutes.

Fourth, risk and uncertainty may play a smaller role in adaptation than mitigation.[8] Adaptation decisions are on a smaller scale, making risk more manageable. There is a greater role for insurance to spread risk. Adaptation measures are generally closer in time to their benefits, and thus uncertainty arising from long time horizons is reduced. And if an uncertain future does in fact turn out to be grim, mitigation is no longer an option, but remedial adaptation can partly reduce damages. Perhaps most importantly, uncertainty concerning catastrophe tilts the balance toward caution and mitigation, as adaptation is inadequate to remedy a widespread ecological and economic collapse. Even so, uncertainty remains important in adaptation decisions. The impact and timing of global warming remains uncertain, complicating adaptation planning. Additionally, those charged with making anticipatory adaptation expenditures are hostage to mitigation targets set in distant and inattentive quarters. Planning a rational adaptation strategy confronts uncertainty concerning mitigation efforts themselves.

Finally, we note the obvious difference. Mitigation, if it had been started earlier and pursued vigorously, would have been capable of avoiding virtually all global-warming damages. Adaptation cannot. The loss of species, the loss of glaciers and coral reefs, the loss of microclimates cannot be reversed. Adaptation can reduce the monetary costs but it is incomplete compensation for these losses. This is not to argue

---

[8] Uncertainty may give an unwarranted preference to adaptation over mitigation investments. Perrings (2003) points out that because uncertainty increases the further we peer into the future, models with high discount rates assign less weight to more uncertain outcomes. The effect is to increase the potential for unexpected future costs but to reduce their present value and weaken the case for mitigation.

that all investments should have been thrown into stabilizing climate, eliminating any adaptation burden. It simply points to another way in which mitigation and adaptation are imperfect substitutes.

Empirical studies connecting mitigation and adaptation approaches to global warming are not common. Tol and Dowlatabadi (2001) and Tol (2004) have examined two areas in which the two approaches interact – vector-borne diseases and sea-level rise – with interesting results. The first study analyzes the effects of emission reductions by OECD countries on the incidence and mortality of malaria and dengue fever in developing countries and illustrates the complexity of the relationship. The analysis models two opposing forces. Global warming increases the range and severity of malaria and other tropical diseases. Emission reductions made by OECD countries would in the first instance reduce mortality levels. But in a global economy, emission reductions in the OECD would reduce export earnings in some developing countries, for example Nigeria with its oil-based economy, and reduce their economic growth. If vulnerability to vector-borne diseases is inversely linked to per-capita income, stronger greenhouse gas abatement in rich countries might lead to a net increase in mortality in poor countries. And indeed they suggest that for the diseases studied, this may in fact be the case. In this perverse situation, mitigation weakens adaptation efforts. This does not imply that emissions reduction flunks a cost-benefit test, as many other benefits are left out. But it does support the idea that the links between mitigation and development, and between development and adaptation, should be part of climate policy analysis.[9]

Tol (2004) again takes up the connection between abatement expenditures and the diversion of resources away from adaptation, but this time in the context of sea-level rise. He avoids the complications of viewing the trade system as transmitting economic-growth shocks. In this analysis, mitigation directly reduces sea-level rise and thereby reduces damages. But the effects of mitigation on sea levels are very sluggish. Holding CO_2 concentrations to 550 ppm would only cut sea-level impacts by 10 percent by 2100. Also, because the costs of mitigation start now, they would divert investments away from economic

---

[9] In a subsequent related analysis, Tol (2008) concludes that an increase in malaria consequent to global warming is unlikely to reverse economic growth in sub-Saharan Africa.

growth. At lower levels of economic activity, countries would be less willing and less able to devote resources to adaptation (that is, protecting coastline by building sea walls). The net result is that mitigation may not significantly diminish the need for adaptation but, due to time lags and competing claims for resources, it may reduce adaptation efforts. The end result of mitigation efforts could be *higher* damages from sea-level rise, at least in the medium term. Once again this result cautions us to avoid considering adaptation and mitigation as simple substitutes, without considering their competition for real resources, and without considering their different time profiles.[10]

Two other studies examining mitigation and adaptation policies and their relation to residual damages are of interest. de Bruin, Dellink and Agrawala (2009) modify a well-known integrated assessment model, DICE, to study the interactions.[11] The modification is to make adaptation expenditures an explicit policy-control variable. To do this, it is necessary to split apart the net damage function in DICE, which had implicitly assumed optimal adaptation expenditures, into two components: adaptation costs and residual damages. Adaptation is treated in a "flow" approach, with expenditures and benefits (damages avoided) occurring in the same time period. This assumption rules out long-term anticipatory adaptation measures. The results of running the adaptation-modified DICE model confirm that indeed mitigation and adaptation are substitutes. The higher the level of one, the lower the optimal level of the other. In the extreme, adaptation is most useful when mitigation is unavailable (the default scenario). At the same time, they are strategic complements in that an optimal policy contains a mixture of mitigation and adaptation expenditures. Because in this model the benefits of mitigation are removed in time from the more immediate costs, whereas adaptation benefits and costs are incurred simultaneously, discounting becomes important. As suggested earlier, the implication is that models with high discount rates will tilt the optimal mix toward adaptation and away from mitigation.[12]

---

[10] Recall also from the previous chapter that Anthoff et al. (2010) found that faster economic growth would be associated with greater protection from sea-level rise.

[11] They also consider the regionally disaggregated sister model, Regional Integrated model for Climate and the Economy (RICE).

[12] There are many reasons to believe that both mitigation and adaptation efforts will fall below their optimum levels. In a companion paper, de Bruin and Dellink (2009)

Bosello et al. (2010) also conclude that mitigation and adaptation are strategic complements. They fold into an optimization model three types of adaptation measures – anticipatory, reactive, and adaptation R&D – to sort out the relative contributions and timing of mitigation and adaptation expenditures in an optimal climate policy. The results appear to depend heavily on two factors: uncertainty regarding a catastrophic outcome, and the discount rate. The lower the discount rate and the greater the concern for catastrophic outcomes, the greater the role for mitigation relative to adaptation. These are not surprising because of the delay between expenditure and damages prevented with mitigation, and because adaptation copes poorly with catastrophic outcomes. They also conclude that expenditures should be tilted toward mitigation in the short run. These analyses are informative, but one should remember we are far from a world in which governments cooperate in a globally efficient mitigation effort, where all adaptation opportunities are recognized and acted on, and where allocations for mitigation and adaptation expenses are drawn from a single purse.

Supply, Demand, and the Green Paradox

Adam Smith famously examined the water-diamond paradox. Why do diamonds, with little intrinsic value, have high prices, whereas water, which is essential, has a low price? His explanation, involving value in exchange versus value in use, and relative scarcity, may need to be revised in light of impending water scarcity. But other paradoxes have arisen to take its place. Diamonds, which are a form of carbon, acquire value when they are mined; carbon, in the form of fossil fuel, is harmless when stored underground but acquires a negative value – a cost – after it is mined and burned. We are willing to pay to bring it to the surface, and then willing to pay again to bury it. A little like monetary gold. The economics of global warming could be seen largely as an effort to keep more carbon *in situ* and less in the atmosphere, or at least slow its relocation.[13] This raises two related and interesting

investigate the effects of barriers to optimal adaptation, including climate change so large and rapid that no adaptation is possible.

[13] Global warming is not the only reason for slowing extraction rates, which may be too high because of insecure property rights for resource owners and open access common property resource features.

strategic questions. What is the optimal time path for the price of carbon? Should policy levers work on its supply as well as its demand?

The dominant approaches to global-warming policy shares two premises; a price should be placed on carbon emissions, and this price should rise over time. Putting a price on emissions would reduce the demand for fossil fuels as conservation opportunities are pursued, and would encourage the development and deployment of alternative-energy sources. The price could be imposed directly, through a carbon tax, or indirectly, as, for example, through a cap-and-trade system. A price of carbon can also be established by creating an "opportunity cost" for emissions as the current Clean Development Mechanism does. And an international system in which developing countries' emissions are "capped" at their business-as-usual (BAU) level and reductions below this level are sold internationally would do much the same thing. These measures all work on decreasing demand for fossil fuels. At the same time, the consensus judgment is that the carbon tax should rise over time. The justifications are that incomes will rise and thus willingness to pay to avoid damages will rise, and, as a stock pollutant, the damage costs of a ton of emissions rise so long as atmospheric concentrations are rising.

The conventional, demand-focused approach – pricing carbon and increasing that price over time – is rather narrow. It is true that taxing a pollutant so that its market price reflects its full social cost has been mainstream economics, if not mainstream policy, since the 1930s. But it has also been known for many years that the rate at which non-renewable resources are extracted and supplied to the market depends on both their current and expected future price. By manipulating price, taxes affect supply as well as demand. These two strands of natural-resource theory were combined in the early 1990s when global warming first attracted significant attention. One question concerned the optimal time path of a carbon tax. The contributors were fully aware that carbon taxes would not only impact demand, but could shift the time profile of supply – an increasing tax, or the anticipation of an increasing tax, could shift forward in time the supply of fossil fuels.[14] A perverse-incentive effect could then result. A carbon tax could front-load the supply of carbon-bearing fossil fuels, depress their price, lead

---

[14] In theory and assuming zero extraction costs, a constant tax would leave the time profile of extraction and supply unaffected.

to higher, not lower, near-term carbon emissions, and could accelerate global warming. This is sometimes known as the "green paradox."[15] The anticipation of increasingly strict carbon controls could also lead suppliers to alter the near-term fuel mix toward fuels with higher carbon content, also accelerating global warming. Smulders and van der Werf (2008) investigate a "dirty-first" fuel-switching model. Their preliminary results suggest that there may be switching from dirty coal to oil, but also from clean gas to oil, a somewhat ambiguous result. Whether the acceleration of global warming is the result of tilting the time path of extraction, or from a "dirty-first" fuel-switching response, rapid warming increases damages by bringing the costs forward in time, and by reducing adaptation opportunities.

The models researching the optimal time path for a carbon tax were inconclusive. Some studies concluded that the optimal carbon tax should decrease over time to prevent premature extraction of fossil fuels and to delay carbon emissions. Some concluded that the time profile of the tax should also reflect increasing damages from a stock pollutant such as CO_2. In that case, the optimal tax may rise and then fall. And one study, which refined the carbon cycle model traditionally used, concluded that the optimal carbon tax could be constant, could increase or decrease monotonically, or could be U- or inverted-U-shaped, all depending on assumptions.[16] Climate IAMs, however, typically show rising carbon prices.

Our second question, the role of supply versus demand in policy response, has received less attention than it deserves. Increasing energy efficiency, subsidizing renewable fuels, gearing up nuclear power, setting auto mileage standards, ramping up R&D for solar or wind or geothermal power, and taxing carbon emissions will be for naught in controlling emissions unless carbon supply reaching the market is curtailed. In that sense, OPEC, at least as much as China, controls the pace of global warming.

[15] We distinguish between shifting demand from one country to another as a result of globally incomplete carbon tax or cap-and-trade systems and shifting the time profile of exploitation. The former is commonly known as carbon leakage and is addressed in Chapter 7. The latter might be termed inter-temporal carbon leakage. Eichner and Pethig (forthcoming) consider the two together. Harstad (2010) analyzes international markets for fossil fuel *deposits* as a supply tool and a response to both types of leakage.

[16] Ulph and Ulph (1994), Sinclair (1994), Farzin and Tahvonen (1996).

Consider, for example, technology policy designed to reduce the cost of renewable energy. Strand (2007) builds a model in which a concerted effort at developing alternatives to fossil fuels may produce a low-cost substitute that renders fossil fuels economically redundant. The time needed to develop such technology is, however, stochastic. Two forces are at work on the fossil fuel extraction rate, and hence the time profile of carbon emissions. The first is that fossil fuel producers will anticipate that the new technology, when it arrives, will shift energy demand to the new, low-cost substitute. Thus they will have an incentive to accelerate extraction in the interim. The second is that when the new low-cost technology arrives and is adopted, extraction of fossil fuels, and hence emissions, cease. These forces are exerted at different points in time and net effect is to shift extraction forward in time.[17] Fossil fuel suppliers also have an incentive to stave off alternative-energy technologies by manipulating their prices downward. In contrast to a technology policy that subsidizes renewable-energy sources, which may accelerate fossil fuel extraction, subsidizing cost-cutting technology improvements in carbon capture and storage may have the opposite effect, encouraging fossil fuel owners to delay extraction in anticipation of lower costs.

The supply response is also relevant in considering the choice between carbon taxes and cap-and-trade systems. If the goal is to slow down and postpone the use of fossil fuels, a uniform carbon tax that increases over time would initially appear attractive for reducing demand and encouraging renewable-energy sources. However, near-term fossil fuel supply also responds to the tax, and according to the logic of the green paradox, this can accelerate extraction. Increased supply is more likely if extraction costs are a small fraction of price (Sinn 2008). The dual response of demand and supply reminds us that tax tools influence but do not set either prices or quantities. In contrast, a universal cap-and-trade system, rigorously enforced, could block the green paradox. A global quantity cap on carbon emissions is just that,

---

[17] Hoel (2009) has also investigated the effects of technology policy to reduce the cost of renewables. He concludes that a successful technology policy could, through demand switching and lower fossil fuel prices, accelerate fossil fuel extraction, increase emissions, and, perversely, increase global temperature. However, this is the result of assuming countries have different carbon taxes, and thus falls within the "carbon leakage" issue discussed in Chapter 7.

a quantity control. The incidence of that cap is a separate question and depends on how the permits are allocated. The allocation may or may not compensate fossil fuel owners for the loss of value of their resources *in situ*. But the owners of the resource cannot protect themselves from a cap that tightens over time by accelerating extraction, as they could with an escalating carbon tax. Instead the global cap could be viewed as a monopsonistic control over demand that invites fossil fuel producers to counter with monopolistic supply controls. The incidence of the losses created by the cap as between producers and consumers is indeterminate, but so long as the global cap remains, extraction, emissions, and temperature all escape the green paradox.

For reasons explored in Chapter 8, a universal cap-and-trade system is unlikely for some time. One proposal being given serious consideration is to set a quantitative cap for rich countries and allow them to meet that commitment by buying carbon reduction credits from developing countries that hold their emissions below BAU levels. This would have the desirable results of shifting abatement effort to the lowest-marginal-cost countries, and provide funds to developing countries to participate in the global effort. But will it resolve the green paradox? The answer depends in part on how BAU emissions are calculated. If BAU emissions projections are based on the "high" and rising fossil fuel prices that would be expected to prevail without a cap-and-trade system, the opportunity cost of emissions reduction will be relatively high and emission reductions in developing countries would be discouraged.[18] The reason is that BAU projections would have factored in low emissions due to high fuel prices, in which case reductions below these levels would be difficult to make. However, if fossil fuel prices are expected to collapse due to the cap adopted by the rich countries, and BAU projections are based on "low" fossil fuel prices, the opportunity cost of emissions reductions is relatively low and reductions would be encouraged. If the global cap is firm, the allocation of emission reduction between rich and poor will depend on how the BAU is set, but the green paradox is avoided. But if the cap itself is sensitive to total costs, flexibility in setting BAU levels can create slack and reintroduce the green paradox problem.

---

[18] That is the BAU assumption is prices rising a la Hotelling at a rate equal to the real interest rate.

One policy measure that would break the physical link of moving carbon from underground, where it is environmentally harmless, to the atmosphere, where it is harmful, is sequestration through afforestation accompanied by reduced deforestation. Another is promoting carbon capture and storage technologies. The tool used in these policies would presumably be subsidies. But if the international community is willing to subsidize these activities, why not attack the problem directly by offering subsidies to resource owners to keep fossil fuels *in situ*? If subsidies are out of the question for political reasons, perhaps some assurance that the owners would not bear a disproportionate share of the real losses associated with climate policies might be useful. Indeed one may take this a step further and create a market for maintaining fossil fuel reserves much like proposals to maintain carbon sequestered in forests.

It is difficult to say how much attention should be paid to the green paradox. The basic theory on which it rests – Hotelling's conclusion that under certain conditions including competitive markets the price of exhaustible natural resources rises at the real rate of interest – has shaky empirical confirmation. And accounting for extraction costs weakens the paradox. Expectations of future carbon taxes are important. If a policy of increasing taxes is not seen as credible, the green paradox evaporates. Moreover, energy resources are owned by a rich mixture of public and private entities, and energy markets are certainly less than fully competitive. Energy companies have or aspire to obtain long-term contractual obligations, and have large fixed assets in downstream transport, distribution, and processing, all of which influence their extraction decisions. Despite these qualifications, however, one should anticipate that demand reduction policies can be dampened by supply responses.

Technology

Technological change is widely considered to be the key to controlling the costs of mitigating global warming, and will be important in cost-effective adaptation.[19] Indeed, success in controlling temperature will require a profound technological transformation of energy

---

[19] For a comprehensive discussion, see Popp, Newell, and Jaffe (2010).

production and use in the next few decades. Can market incentives be manipulated to support this transformation? How is the global dissemination of these technologies to be financed? Incorporating technology into global-warming analysis and finding effective technology policies has been difficult. One reason is that the economics of technological change has no overarching theoretical structure. Another is that data for empirical tests are often deficient. These difficulties are partly the result of technology's heterogeneous nature and the multiple points at which it interacts with climate change: improvements in energy efficiency that reduce emissions; substitutions among fossil fuels with different carbon content; opportunities for and cost of alternative-energy sources such as hydro, solar, geothermal, wind, and nuclear; the prospects for and cost of carbon capture and storage; the prospects for breakthrough technologies (nuclear fusion?); and a variety of geo-engineering proposals to alter the earth's energy balance. At a more fundamental level, the socioeconomic drivers for technological change and its widespread deployment are not clearly understood.

Is Technology Policy Needed?

Perhaps the most basic question is whether an explicit technology policy is needed. In the early literature technological change was often treated as an exogenous time variable, sometimes with a specific energy-efficiency parameter that contributed to decreasing emissions intensity of output. Rising fossil fuel prices (Hotelling's Rule again), coupled with assumed decreasing costs for non-carbon backup energy sources such as solar photovoltaic technology, could eventually displace fossil fuels and bring down emissions.

Exogenous technological change is not very satisfactory on either theoretical or policy grounds. Theory suggests that innovation and deployment of new technology is subject to various market imperfections, and especially the failure of innovators to capture the full social benefits of their activity. If this is correct, historical data on technological change may reflect the imperfections, not the true potential of technology to reduce emissions cost. The alleged imperfections in the introduction of new and more efficient technology boil down to four: the failure until now to internalize the externality cost of carbon emissions, and hence the lack of an incentive to develop carbon-sparing technology; the inability to appropriate to the innovator the full social benefits of her activities; deficiencies in private capital markets that may prevent funding high-cost but risky new technologies; and barriers to "learning-by-doing" cost reductions.[20] The theory of policy, which dates to Tinbergen (1956), concludes that governments need at least as many policy instruments as there are market correction objectives. This implies that a single policy of setting a carbon price by tax or cap-and-trade is insufficient, as it addresses the external costs of carbon emissions and only indirectly affects technology market imperfections.[21] By symmetry, it suggests that subsidies can addresses environmental technology market imperfections, but fail to directly address externalities from emissions.[22]

The limits of assuming exogenous technological change are clear, and models of policy-induced technology change have been constructed.[23] Typically, three policy channels are identified: The first explores the indirect effect of carbon pricing on the direction and pace of technological change in the energy sector, and the second and third examine direct technology policies – support for R&D and support for rapid diffusion and adoption of new technologies. The last assumes that the unit cost of using a new technology is a decreasing function of the cumulative output using that technology. The first, carbon pricing, affects the relative price of energy, an input to production, and sometimes a direct consumption good (e.g., household heating and cooling). The price increase encourages substitution away from energy-intensive production and consumption, but this is not itself technological change as conventionally defined. But higher fossil-fuel-based

---

[20] Learning-by-doing theory bears a remarkable similarity to the 200-year-old infant industry argument for trade protection. For example, from 2003 to 2009, China – now a powerhouse in wind turbines – required its wind farms to use locally made turbine parts to build output and cut costs. U.S. senators have suggested a border tax on U.S. imports of clean energy technology, presumably for the same purpose.

[21] Keep in mind that the purpose of carbon pricing policy is to correct a negative externality (external diseconomy) and the purpose of technology policy is to correct for a positive externality, inadequate private investment in new technology.

[22] This does not mean emissions policy and technology policy are equally effective. Popp (2006) calculates that an optimal emissions tax alone could capture 95 percent of the welfare benefits of a combined emissions and technology R&D policy, whereas the optimal R&D policy alone would only capture 11 percent of the combined welfare benefits.

[23] Technological change occurs in the absence of explicit policy measures. It is difficult to separate autonomous from policy-induced change.

energy prices also stimulate R&D in energy-saving technology and renewable-energy sources. Additionally, higher prices stimulate output and thus learning-by-doing cost reductions in renewable-energy production. Carbon pricing does not, however, directly address defective incentives for technological improvements and deployment.[24]

Knowledge has long been recognized as exhibiting classic public good characteristics – non-exhaustibility and imperfect excludability – and deserving of public support. The social returns to innovation are thought to exceed the private returns. Support can be extended through tightening intellectual property protection (e.g., patents), government R&D, or private-sector R&D subsidies and tax breaks. These policies can be general or specific to energy. Both can be thought of as supplemental to carbon pricing. But support for innovation may not be enough. A combination of scale economies, information deficiencies, and inertia can slow the introduction of new technologies. And accelerating the introduction can capture network externalities associated with expanded use of the new technologies. Subsidies, technology mandates such as renewable-fuel standards, electric-car mandates, energy-efficiency standards for appliances, and other policy tools are used to accelerate adoption of new technologies and capture learning-by-doing cost reductions. A credible commitment to maintaining these policies can help. Demonstration projects can also play a role.

Mitigation-Technology Connections

In a previous section, we explored the interaction between mitigation and adaptation policies. There are also interactions between mitigation and technology policies. In terms of externality theory, the external diseconomy of emissions should be dealt with through a carbon tax, and the public-goods-type positive externalities of technology should be corrected with a second set of policies. In the absence of direct technology policies, however, we are in a second-best world, and a case can be made that carbon taxes should be increased to indirectly promote technological improvements.[25]

---

[24] Different carbon pricing tools such as taxes, auctioned and non-auctioned tradable permits, and abatement subsidies can have different impacts on technological change incentives. Milliman and Prince (1989) find that taxes and auctioned permits provide firms the greatest incentive.

[25] It is unlikely that a single policy – carbon taxes – can obtain the first best welfare level that could be obtained if both policies were available.

Assuming both policies are available, directly promoting technology should lower emissions abatement costs and, in a BC calculus, justify greater mitigation and a more aggressive global-warming policy. At first glance, it would also appear that the optimal carbon tax should be lower, as some of the burden of abatement is shouldered through cost-cutting technology policy. However, even the seemingly straightforward conclusion that technology policy supports a more stringent emission abatement policy can be challenged. The conventional view that technological change lowers the total and marginal emissions abatement costs may be incorrect. Total abatement costs may fall but marginal abatement cost curves become steeper at high abatement levels (Baker et al. 2008). Baker argues that this is more likely when technologies that are appropriate at low levels of abatement give way to more radical technologies at higher levels of abatement. If technological change lowers total abatement costs but increases marginal abatement costs, the result is a lower optimal level of abatement (higher emissions) and a higher emissions tax.

Another channel through which emissions policy and technology policy interact is the timing of policy. If one believes that R&D is the principal vehicle through which abatement costs can be cut, and if a large-scale R&D program is mounted, there may be reason to slow down emissions abatement targets until such time as abatement costs have fallen. But if direct R&D efforts are anemic, the burden of inducing technological change falls more heavily on emissions policy and supports setting a high carbon price to *indirectly* stimulate R&D. Gerlagh et al. (2009) find that in this case, emissions taxes should be set high relative to their Pigouvian level.[26] The conclusions are less clear if learning by doing is the main vehicle through which abatement costs cuts are to be achieved. In this situation, Goulder and Mathai (2000) find the impact of induced technological change on the optimal timing of abatement to be ambiguous.

Empirical Studies

Considerable empirical work links technology and mitigation costs. We offer a few samples. An early study by Chakravorty et al. (1997)

---

[26] A Pigouvian tax would be set at a rate equal to the externality cost at the socially optimal level of greenhouse gas emissions. Hart (2008) also considers the timing of emissions taxes when technological change is endogenous.

examined the opportunity to displace fossil fuels with a backup technology (solar voltaic) under different technological change assumptions. The simulation model accounted for multiple fuels (coal, oil, and natural gas) mated to different demands (transport, heating, electricity, etc.). The optimistic conclusion was that if solar energy could continue its historic rate of cost reduction, carbon emissions would peak by 2030 and temperature increase would be limited to about 1.5°C. Some 90 percent of world coal would be left safely underground. With a somewhat less optimistic technology assumption, roughly the same result could be obtained if in addition a $100 per ton carbon tax were employed to motivate the replacement of fossil fuels.[27]

Popp (2002) is concerned with a specific question – the effect of energy prices on energy innovation – but is also interested in the more fundamental question of the roles of demand and supply for innovation. The demand presumably comes from a change in energy prices that raises the return on an energy-efficient innovation. Popp views the supply being influenced by the stock of useful knowledge on which an energy-saving innovation is built. His approach is regression analysis using energy price data and U.S. energy-related patents for 1970–1994. When interpreted in the context of global warming, his results tend to confirm that pricing carbon can have a strong impact on energy sector innovation and hence cost savings. But a correct understanding of induced innovation also requires appreciation of the role played by stock of useful knowledge.

The nexus between carbon pricing, technological progress, and mitigation costs has also been explored with World Induced Technical Change Hybrid (WITCH), an integrated assessment model featuring endogenous technology (Bosetti et al. 2009). The two channels for technological progress in carbon emission mitigation – R&D and learning by doing – are explicitly modeled. Incremental improvements in low-carbon and renewable-energy technologies are considered separately from breakthrough, zero-carbon, "backstop" technologies. Among the more interesting results, the model shows that current and expected future carbon prices can have a strong impact on R&D spending and on the speed with which clean technology is diffused. But setting aside

---

[27] The costs of converting solar energy to electricity were assumed to fall by 50 percent and 30 percent per decade in the two cases.

breakthrough technologies, the impact of these incremental technological improvements on mitigation costs is very minimal. No silver
bullets here. The reason appears to be that many incremental technological innovations are already with us today or are in train, and
only need implementing. The barriers include currently uncompetitive
prices and restrictions on deployment, for example nuclear power. This
is indirectly confirmed by a simulation that freezes nuclear energy at
current levels, rules out carbon capture and storage, and caps wind and
solar at 35 percent total electricity production. Under these restrictive
circumstances, mitigation policy costs rise from 3.9 percent to 7 percent of gross world product in 2050.

In this model, the contrast between investment in incremental and
in breakthrough technologies could hardly be greater. This model calculates that to achieve a greenhouse gas concentration of 550 ppm –
an ambitious and much discussed objective – the development and
deployment of a "breakthrough" technology could cut mitigation
costs from 4 to 2 percent of gross world output in 2052 and from 7.6
to 1.4 percent in 2082.[28] Even here, however, there are no free lunches.
Carbon prices would need to rise in the short and medium term to
provide R&D incentives. And even though breakthrough technologies can generate long-term cost savings, they impose welfare losses in
the short and medium term.

Geo-Engineering

Geo-engineering can be defined as deliberate interventions in the
global climate system to moderate or prevent global warming. Until
recently very much on the fringe, geo-engineering proposals recently
have gained some credibility and respectability. That new status
has less to do with new and favorable research results and more
with growing pessimism that an effective mitigation or mitigation-
plus-adaptation response will be mounted. The tone of geo-engineering
discussions often has echoes of retreat or defeat, even among many
proponents.

---

[28] In this analysis, nuclear power is not considered a breakthrough technology, but its
deployment is severely limited. Without this limit, there would be little need for a
breakthrough technology in the electricity sector.

The proposals themselves are diverse and display considerable inventiveness. Broadly speaking, they fall into two categories: increasing terrestrial and ocean uptake of greenhouse gases, and altering the earth's radiation balance by reducing incoming radiation or increasing its reflectivity (albedo). An example of the first is "fertilizing" the oceans' surface with phosphate or iron to stimulate algae growth and the absorption of atmospheric carbon. Examples of the second are space-based sunlight deflectors, and dispersing sulfate particles into the stratosphere to increase reflectivity. The last has gained the most attention.

The science, technology, economics, and politics of these proposals are still in their infancy. At this point, it may be most useful to simply list some of the potential advantages and disadvantages. One purported advantage is that certain technologies, and especially dispersing stratospheric sulfate particles, would appear to be far less expensive than mitigation.[29] Net benefits – mitigation and adaptation costs avoided minus geo-engineering costs – could be in the trillions of dollars. A second advantage over mitigation is that it could sidestep the problem of funding a global public good. If in fact costs are as low as touted, the national benefits of unilateral action may easily cover the costs, and the global public goods dilemma that is hanging up mitigation efforts is finessed. One country acting in its own self-interest could justify the expenditure. A third purported advantage over mitigation is that if preparatory research and testing has been done, some technologies can be deployed quickly with almost immediate effect. The eruption of Mt. Pinatubo in 1991, which is frequently cited as a quasi-natural experiment in atmospheric sulfate particles, had measurable global cooling effects within six months (Crutzen 2006). The rapid action suggests that geo-engineering through sulfate particles may be particularly attractive as a backup option if mitigation fails or if runaway temperature increases are imminent.

These are potentially strong advantages. But one can list potentially strong disadvantages as well. First, serious adverse side effects may exist. For example, dispersal of sulfate particles in the atmosphere may control temperature but would do nothing to limit acidification of the oceans and problems related thereto. Depletion of atmospheric ozone

[29] For evaluation of geo-engineering from different perspectives, see Barrett (2008); Curtzen (2006); Teller et al. (2003); Victor (2008).

and disruption of local climates may also result. There are known and unknown risks.

Second, if geo-engineering solutions come to displace mitigation as the dominant strategy, the mitigation option is lost. CO_2 would have in the meantime accumulated, and the geo-engineering solution would have to be continued to avoid a disastrous, rapid run-up in global temperatures. Geo-engineering may be able to substitute if we fail at mitigation, but if geo-engineering fails, mitigation is no longer an option. This is not a trivial point. Geo-engineering as a backup is attractive, but once researched and tested the temptation to promote it over the more expensive mitigation may be overwhelming. To put the point a little differently, mitigation, with its high cost, vulnerability to free-riding, and benefits in the far future, is a tough sell. The prospect of a geo-engineering alternative might make it an impossible sale. But later on, a nasty surprise with the geo-engineering solution would leave the policy tool chest quite empty. Even if there is no formal public decision to switch from mitigation to geo-engineering, the prospect of the latter undermines the credibility of increasing carbon prices in a mitigation strategy. That, in turn, can have potentially severe effects on investment and R&D decisions needed for successful mitigation.

A third consideration is the prospect of international political conflicts using geo-engineering tools. Divergent environmental, economic, and political interests are certainly obstructing the mitigation option today. But they will not automatically disappear when some or many countries have the technology and the resources to manipulate the global climate. On the contrary, geo-engineering may invite action at the national level that has adverse effects on others, including drought and monsoon rain patterns. Substituting geo-engineering climate externalities for international emissions externalities would hardly be a step forward. Serious consideration of geo-engineering technologies needs to be accompanied by serious consideration of international governance issues (Victor 2008).

Fourth, reliance on geo-engineering over mitigation would represent an ethically questionable shifting of risk to future generations.

Conclusions

Chapters 2–4 concentrated on the economics of mitigation (abatement) within a BC framework. This chapter expands that perspective by considering other responses to global warming: accelerated economic development; adaptation; controlling carbon supply as well as demand; technology; and geo-engineering. These strategies are not mutually exclusive. The challenge is to find the appropriate balance. The principal conclusion is that mitigation needs to remain the centerpiece, although each of the others can make a contribution.

Accelerated economic development in poor countries is desirable in itself. But apart from its value in improving adaptive capacity, it makes no direct contribution to solving the root of the global-warming problem, which is inter-temporal externality costs of uncontrolled greenhouse gas emissions. On the contrary, rapid economic development is a key driver in emissions growth. The important task is to control emissions without penalizing growth in poor countries.

Adaptation can be a valuable supplement to mitigation, but the two are imperfect substitutes at best. Efficient adaptation can reduce optimal mitigation costs just as efficient mitigation can help reduce adaptation costs. Still, adaptation does nothing to solve the emissions externality problem. Unlike mitigation, it does nothing to minimize catastrophic risk, although it may reduce damages. It would be unfortunate if international funding for adaptation were at the expense of funding mitigation.

The mitigation policy literature is dominated by measures to control emissions by controlling demand. The green paradox reminds us that there may be a perverse supply response for non-renewable resources such as fossil fuels. This insight does not give rise to a totally new approach to global warming, but it does cast some light on the appropriate policy mix. Specifically it tends to support quantity over price measures, technologies for carbon capture and storage, sequestration through afforestation, and, in principle, subsidies to keep fossil fuels below ground.

There are tight connections between mitigation and technology. Mitigation costs will depend on technological progress. Technological progress depends in large part on price signals arising from mitigation policy. Technology can also be nudged through direct R&D and deployment support, seeking learning-by-doing cost savings. But there is no viable stand-alone technology policy for global warming in the absence of mitigation efforts that provide the correct price signals.

In the future, geo-engineering will occupy a larger space in discussions of the global-warming response portfolio. At the present time, with the mitigation window still open, and with very little serious work on geo-engineering accomplished, there is no case for the latter to displace the former. The critical question is how much effort should be put into understanding geo-engineering solutions better. Doing so may create a powerful temptation to promote geo-engineering from a backup strategy to the principal response to global warming. The attractive chimera is that we could then dispense with all those unpleasant and costly problems associated with mitigation, especially finding the necessary international cooperation. But if we do so, the window for serious mitigation may close, and we are left with a single risky and unproven response. On the other hand, if we do *not* research geo-engineering thoroughly, we may fumble on mitigation and possibly be caught in a catastrophic runaway climate with no shelter. A Hobbesian choice indeed.

References

Anthoff, D., R. Nicholls, and R. S. J. Tol (2010). The Economic Impact of Sea-level Rise. *Mitigation and Adaptation Strategies for Global Change* 15 (40): 321–35.

Baker, E., L. Clarke, and E. Shittu (2008). Technical Change and the Marginal Cost of Abatement. *Energy Economics* 30: 2799–816.

Barrett, S. (2008). The Incredible Economics of Geoengineering. *Environment and Resource Economics* 39: 45–54.

Bosello, F., C. Carraro, and E. De Cian (2010). Climate Policy and the Optimal Balance between Mitigation, Adaptation, and Unavoided Damages. *University of Venice Department of Economics Working Paper* 09/WP/2010.

Bosetti, V., C. Carraro, R. Duval, A. Sgobbi, and M. Tavoni (2009). The Role of R&D and Technology Diffusion in Climate Mitigation. *FEEM Nota di Lavoro* 14.2009.

Chakravorty, U., J. Roumasset, and K. Tse (1997). Endogenous Substitution among Energy Resources and Global Warming. *The Journal of Political Economy* 105 (6): 1201–34.

Crutzen, P. J. (2006). Albedo Enhancement by Stratospheric Sulfur Injections: A Contribution to Resolve a Policy Dilemma? *Climatic Change* 77: 211–19.

de Bruin, K. and R. Dellink (2011). How Harmful are Adaptation Restrictions? *Global Environmental Change* 21 (1): 34–45.

de Bruin, K., R. Dellink, and S. Agrawala (2009). Economic Aspects of Adaptation to Climate Change: Integrated Assessment Modelling of Adaptation Benefits and Costs. *OECD Environment Working Paper No. 6.*

Eichner, T. and R. Pethig (forthcoming). Carbon Leakage, the Green Paradox and Perfect Futures Markets. *forthcoming in International Economic Review.*

Farzin, Y. H. and O. Tahvonen (1996). Global Carbon Cycle and the Optimal Time Path of a Carbon Tax. *Oxford Economic Papers* 48: 515–36.

Gerlagh, R., S. Kverndokk, and K. E. Rosendahl (2009). Optimal Timing of Climate Change Policy: Interaction between Carbon Taxes and Innovation Externalities. *Environment and Resources Economics* 43: 369–90.

Goulder, L. and K. Mathai (2000). Optimal CO_2 Abatement in the Presence of Technological Change. *Journal of Environmental Economics and Management* 39: 1–38.

Harstad, B. (2010). Buy Coal? Deposit Markets Prevent Carbon Leakage. *NBER WP 16119.*

Hart, R. (2008). The Timing of Taxes on CO_2 When Technological Change is Endogenous. *Journal of Environmental Economics and Management* 55: 194–212.

Hoel, M. (2009). Bush Meets Hotelling: Effects of Improved Renewable Energy Technology on Greenhouse Gas Emissions. *Fondazione Eni Enrico Nota di Lavoro* 01.2009.

Milliman, S. and R. Prince (1989). Firm Incentives to Promote Technological Change in Pollution Control. *Journal of Environmental Economics and Management* 17: 247–65.

Nordhaus, W. (2008). *A Question of Balance: Weighing the Options on Global Warming Policy.* New Haven CT: Yale University Press. Prepublication version at http://www.nordhaus.econ.yale.edu/Balance_prepub.pdf

Perrings, C. (2003). The Economics of Abrupt Climate Change. *Phil. Trans, R. Soc. London* 361: 2043–59.

Popp, D. (2002). Induced Innovation and Energy Prices. *American Economic Review* 92 (1): 160–80.

    (2006). R&D Subsidies and Climate Policy: Is There a "Free Lunch"? *Climatic Change* 77: 311–41.

Popp, D., R. Newell, and A. Jaffe (2010). Energy, the Environment, and Technological Change in *Handbook of the Economics of Innovation: Vol 2.*, B. Hall and N Rosenberg (eds.). Academic Press/Elsevier.

Sinclair, P. (1994). On the Optimal Trend of Fossil Fuel Taxation. *Oxford Economic Papers* 46: 869–77.

Sinn, H-W. (2008). Public Policies against Global Warming: a Supply Side Approach. *International Tax and Public Finance* 15 (4): 360–94.

Smulders, S. and E. van der Werf (2008). Climate Policy and the Optimal Extraction of High- and Low-Carbon Fossil Fuels. *Canadian Journal of Economics* 41 (4): 1421–44.

Stehr, N. and H. von Storch (2005). Introduction to Papers on Mitigation and Adaptation strategies for Climate Change: Protecting Nature from Society or Protecting Society from Nature. *Environmental Science and Policy* 8: 537–40.

Strand, J. (2007). Technology Treaties and Fossil-Fuel Extraction. *The Energy Journal* 28 (4): 129–41.

Teller, E. et al. (2003). Active Stabilization of Climate: Inexpensive, Low Risk, Near-term Options for Preventing Global Warming and Ice Ages via Technologically Varied Solar Radiative Forcing. Lawrence Livermore National Library, 30 November.

Tinbergen, J. (1956). *Economic Policy: Principles and Design.* Amsterdam: North Holland.

Tol, R. S. J. (2004). The Double Trade-off Between Adaptation and Mitigation for Sea Level Rise: an Application of FUND. *FNU Working Paper 48.*

—— (2008). Climate, Development, and Malaria: an Application of FUND. *Climatic Change* 88: 21–34.

Tol, R. S. J. and H. Dowlatabadi (2001). Vector-borne Disease, Development, and Climate Change. *Integrated Assessment* 2: 173–81.

Tol, R. S. J., S. Fankhauser, and J. B. Smith (1998). The Scope for Adaptation to Climate Change: What Can We Learn from the Impact Literature? *Global Environmental Change* 8 (2): 109–23.

Tulkens, H. and V. Van Steenberghe (2009). "Mitigation, Adaptation, Suffering": In Search of the Right Mix in the Face of Climate Change. *Fondazione Eni Enrico Nota di Lavoro 79.2009.*

Ulph, A. and D. Ulph (1994). The Optimal Time Path of a Carbon Tax. *Oxford Economic Papers* 46: 857–68.

Victor, D. (2008). On the Regulation of Geoengineering. *Oxford Review of Economic Policy* 24 (2): 322–36.

6

Targets and Tools

The policy process typically identifies objectives, establishes targets, and selects tools or instruments to attain the targets. This chapter is mainly concerned with the tools for achieving climate policy objectives and targets. This involves a discussion of the putative advantages of market-friendly tools that use incentives and disincentives to limit greenhouse gas emissions versus regulatory tools. It also involves evaluating the three main types of market-oriented tools: taxes, cap-and-trade schemes, and subsidies. Before starting this discussion, however, it is useful to compare two types of emissions targets. As described in Chapter 9, both types are found in the Copenhagen Accord commitments and are causing some confusion.

Absolute versus Intensity Targets

Emissions of greenhouse gases are the control variable in climate policy. Emissions *targets* can be set in absolute quantities or in relative terms. Absolute targets are expressed in tons of CO_2e and can be set at the sector, national, or global level. Relative targets are commonly known as intensity targets and are set at either the sector or national level. We are mainly interested in national-level absolute targets and in national-level intensity targets expressed per unit GDP – that is, tons of emissions per million dollars of GDP.[1]

---

[1] For a proposal to set international, sector-specific intensity targets, see Schmidt et al. (2008). Most intensity target discussions are limited to CO_2. Data on other gases are less complete so carbon intensity of GDP is the appropriate term. As of 2010, the United States had no formal absolute target, but the Obama Administration has

Certainty of GDP Growth

If future GDP were known with certainty, absolute and intensity targets would be perfectly convertible. An absolute emissions target divided by the target-year GDP is the equivalent intensity target. An intensity target multiplied by the (known) future GDP is the equivalent absolute target. Note, however, that even when GDP growth is accurately projected, if GDP growth exceeds emission growth, emissions *intensity* will be decreasing while *emissions* are still growing.

With certainty concerning GDP, absolute and intensity targets are interchangeable, but the common perception of them is often quite different. There are three reasons. First, for most countries, the carbon intensity of GDP is declining without any explicit climate policy. This is the result of structural shifts in output away from carbon-intensive heavy industry, fuel switching, and increased efficiency in the use of energy. These numbers can be quite dramatic. Between 1980 and 2004, China's carbon intensity declined by 63 percent and the U.S. intensity declined by 40 percent despite increases in absolute emissions in both countries (Timilsina 2008). These exogenous intensity trends can be expected to continue, and it is rather easy to confuse projected *autonomous* declines in carbon intensity with *policy-induced* declines. For example, after rejecting the Kyoto Protocol, President Bush proposed the goal of an 18 percent emissions *intensity* reduction for the decade 2002–2012. Although this may appear to some as impressive, it was criticized by others as little more than business as usual (BAU) and a subterfuge for avoiding serious action.[2]

Second, whereas intensity targets can piggyback on exogenous trends and give the appearance of quite striking progress, emissions targets set in absolute terms have a tougher sell. Even with a very serious climate policy, emissions will rise over the next two decades in most countries. Using absolute targets, emissions will be seen as rising, whereas the equivalent intensity target will show declining intensity. It may be easier politics to market a plan for declining emissions

announced a target of 17 percent below 2005 levels by 2020, subject to passing domestic legislation. Post-Copenhagen China and India announced aspirational targets for 2020 of reducing carbon intensity by 40–45 percent and 20–25 percent, respectively.

[2] The EPA has estimated that the CO_2 intensity of U.S. production of six energy-intensive sectors will decline by about 20 percent between 2006 and 2020 even in the absence of any climate legislation.

intensity than a plan that merely slows the increase of emissions, even when the amount and trajectory of emissions are the same.[3]

A third reason why intensity targets may appear attractive is the perception that they do not limit economic growth, or are less onerous for fast-growing economies. This is incorrect in a certain world and brings up an important point. Intensity targets are not inherently less stringent than absolute targets. Absolute and intensity targets designed to produce the same level of emissions will have the same cost and the same impact on growth – minimal if the targets are minimal, and larger with stricter targets. Confusion arises because an intensity target allows emissions to increase with economic growth. However, an absolute target, set to achieve the same future emissions level as the intensity target, will also accommodate economic growth.[4]

Uncertain GDP Growth

The interchangeability of absolute and intensity targets breaks down in a world where GDP growth is uncertain, but it is not clear whether uncertainty favors one approach over the other. Uncertainty in future GDP means uncertainty in carbon intensity and abatement costs with an absolute target, but it means uncertainty in emissions if an intensity target is chosen. In part this difference parallels the price versus quantity uncertainty issue discussed later in the chapter.

Suppose an absolute emissions target is chosen on the basis of an expected GDP and GDP turns out to be 20 percent higher than expected. Realized emissions are unaffected, but abatement costs are unexpectedly high. Had an equivalent intensity target been chosen, realized emissions would be higher than they would have been under the absolute target, but abatement costs would be lower. The absolute

---

[3] For example, if world GDP were to grow by 30 percent over the next decade, a target of 16 percent reduction in carbon intensity is identical to an absolute target allowing a 9.2 percent increase in emissions. It may be harder to generate enthusiasm for the latter. See Pizer (2005) on this point.

[4] For example, if current emissions are 100, GDP is 1,000, and annual growth is known to be 10 percent, an absolute emissions target of 80 one year hence requires an intensity target of .0728. Both targets require emissions to fall to 80 from the level they otherwise would have risen to. Now assume instead that growth is known to be 2 percent. With the same absolute emissions target of 80, the intensity target is more relaxed at .0785. While the cost of achieving the targets is presumably lower as less abatement is necessary, the costs are the same under the absolute and intensity approaches.

target stabilizes emissions; the intensity target stabilizes costs. If GDP turns out to be 20 percent lower than expected, realized emissions under an absolute target would be unaffected. Abatement costs would be unexpectedly low, or zero. Had an intensity target been chosen, realized emissions would have been lower than under the absolute target, but abatement costs would have been higher. The absolute target again stabilizes emissions. The intensity target again stabilizes costs. The intensity approach may sound attractive, as emissions are allow to rise *pari passu* with output, but they must also fall when times are tough. They appear flexible with an unexpected boom but inflexible in a recession. There are economic and political economy consequences to the choice between absolute and intensity targets.

Ellerman and Sue Wing (2003) illustrate these differences by simulating emission levels and abatement costs in the European Union under both types of targets when GDP surpasses or falls short of expected levels. The analysis shows that emissions are invariant across unexpectedly high and low growth under an absolute target, but are almost 50 percent higher in an unexpectedly high growth outcome than in an unexpectedly low growth outcome under an intensity target. The costs of abatement, measured as a percent of GDP, range from 2.5 percent to 0 for an absolute target but show a more modest range of 1.2–0.9 percent when an intensity target is employed.

A fair comparison requires the two targets to be interchangeable at the expected GDP and then recognizes that actual GDP will fluctuate around expected GDP. The superiority of one approach over the other requires comparing the opportunity cost of abatement when actual GDP departs from expected GDP. Under an absolute target, the amount of *abatement* increases in a boom. Under an intensity target, the amount of *abatement* also increases in a boom but by a smaller amount. Under an absolute target, abatement falls in a recession. Under an intensity target, in a recession, abatement also falls but by a smaller amount. The question then is whether the resources used for abatement have a higher social value (*opportunity cost*) in a boom, in which case the intensity target has the edge, or whether resources used for abatement have a higher opportunity cost in a recession, in which case the absolute target gains the edge. Karp and Zhao (2009) describe a general-equilibrium model showing that with unexpected booms and busts, the intensity target leads to a higher level of welfare but creates greater variability in income and welfare. A risk-averse country may still prefer the absolute target.

Choices

The political economy argument concerning target choice is in two parts. One strand stresses that there is a strong concern in rich and poor countries that mitigation will limit economic growth. Attention is focused on whether targets could choke off an economic boom. In this regard, an unexpected boom would favor an intensity target. Much less attention has been given to the possibility of unexpected slow growth, in which case increases in abatement obligations under an intensity target could prolong and deepen a recession. The second strand of the political economy argument is that all targets are vulnerable to reneging, but intensity targets require *greater* abatement the more severe a recession and thus are more likely to be scrapped in a downturn. But this also is a blinkered view – absolute targets may be more likely to be scrapped than intensity targets in a boom. Unfortunately, one cannot have absolute targets when GDP is exceptionally weak and intensity targets in an unexpected upturn, and still maintain a credible climate policy.

No clear conclusion emerges. Countries primarily concerned with their growth prospects may chose intensity targets either through a cynical calculation that such targets can give the illusion of serious action, or because they are giving unequal weight to unexpected booms versus unexpected slowdowns. Remember, if GDP growth was known with certainty, absolute and intensity targets would be interchangeable. Countries that take global warming very seriously make take more comfort in absolute targets, but they should at least be aware that on a global level, those targets will increase over the next two decades, and that with certain GDP growth, the same emission levels could be obtained with intensity targets.

Two final points that weigh against intensity targets might be made. First, a proper international comparison of intensity targets requires agreement on two variables: emissions and GDP. Deflating nominal GDP and choosing the appropriate exchange rate to convert to a common currency to make international comparisons introduce areas of possible dispute, not present in absolute standards. Second, setting intensity targets for the economy invite (but do not require) intensity standards sector by sector. This step may create unwelcome complications such as unproductive rent seeking at the sector level and manipulating sector output levels to gain higher emission limits. To be fair, however, setting an absolute national emission target does not rule out using sector intensity targets.[5]

The Toolbox

The challenge of climate change has not produced any radical breakthrough in environmental policy and policy analysis.[6] However, it has enormously increased the saliency of two earlier insights: the advantages of market-based tools over so-called command and control, or prescriptive, policies: and the related issue of regulating through price or quantity. After the almost endless hums and haws in the global warming economics literature, it is refreshing to find nearly universal agreement among economists on one policy issue. Market-based tools that put a price on greenhouse gas emissions should be the centerpiece. And although there continues to be disagreement on price-versus-quantity measures, there is widespread agreement on a second issue. If a cap-and-trade scheme is chosen, to the extent possible, emission permits should be auctioned off and not given out for free.

Broadly speaking, the environmental policy toolbox has three compartments. Policies in the first compartment set specific standards in terms of inputs, technology, performance, or emissions. These are sometimes known as command-and-control or regulatory measures. Mandates for minimum energy from renewable sources in electricity production (so-called portfolio standards), mileage standards for autos, and point source effluent standards are examples. There are situations in which these prescriptive policies are efficient or relatively easy to

---

[5] Following the 2009 Copenhagen meeting, South Korea joined China and India with an intensity pledge; the EU and Japan pledged absolute reductions. Some other countries have pledged reductions below BAU levels, but the effective reductions cannot be known until there is some agreement on BAU levels.

[6] The following sections primarily address carbon emissions from fossil fuel combustion. Other factors including land use changes (deforestation) and carbon sequestration, are also important, as are other greenhouse gases. A full consideration encounters additional issues: establishing a counterfactual baseline for biological sequestration; monitoring non-point source emissions; and making different greenhouse gases commensurate.

monitor and enforce. Bans or severe restrictions on dangerous pesticides and a requirement for double hulls for oil tankers are examples. However, they are rarely efficient or effective for global warming.

A second compartment, further to the right, contains a single tool. Government clarifies private property rights to the environment, thus allowing the establishment of a market and letting environmental goods and their quality be determined in that market by transactions between polluters and victims. This follows from the Coase Theorem, which demonstrates that under certain restrictive assumptions, the resulting allocation of resources will be efficient regardless of how the property rights are allocated. Unfortunately, global warming fails to meet the basic conditions for an efficient Coasian market. The global atmosphere resists privatization, and because of time lags, no market between this generation and generations yet to come can be established. Moreover, the sheer number of participants would generate insoluble free-rider problems.[7]

The third compartment contains economic instruments that use market incentives, and is of greatest interest. These policy tools are emissions taxes, cap-and-trade (tradable permit) schemes, subsidies for pollution reduction below a benchmark, and subsidies to encourage green technology.[8] As shown later in the chapter, they are especially well suited to the problem of global warming.

Government policy and tools can be evaluated from a number of perspectives. These include efficiency (reducing emissions at least cost), equity (distributional effects), response to uncertainty and risk, flexibility in accommodating new information, fiscal effects, administrative and monitoring requirements, political acceptance, and others. There are often trade-offs among the criteria, and several tools may be employed simultaneously. Economics has most to say about the efficiency of policy but does not totally neglect other dimensions.[9]

---

[7] The economic tools discussed later in the chapter do follow Coase in establishing emission rights, but the rights are initially held by the government. In cap-and-trade systems, the rights are acquired by emitters from governments and are subsequently traded among themselves.

[8] There are other tools including energy conservation through zoning, provision of information, demonstration projects, and moral suasion. All may have some role in a comprehensive climate policy.

[9] For a useful discussion, see Goulder and Parry (2008).

Market Incentives versus Regulation

Three features of the greenhouse problem tend to favor market-incentive approaches. First, the gases become uniformly mixed in the atmosphere so that each unit carries the same potential for damage as every other unit. It follows that reducing emissions by one ton *from any source* is equally desirable. Additionally, for efficiency, the marginal cost of abatement should be equal for all sources. A single homogenous "product" – a ton of carbon abatement – and a single uniform efficient price for that product result.[10] This contrasts with many other pollutants whose damage potential may depend on where and at what level of dilution they are emitted.[11] In the latter case, efficiency would require multiple "products," conversion factors, and multiple prices.

Second, there is wide variation in the marginal cost of greenhouse gas abatement among activities, sectors, and countries. Pricing carbon emissions through economic instruments lets the market sort out the amount of abatement undertaken by different sources. Each source will compare its marginal abatement cost to the price of a ton of emissions, and will undertake abatement up to the point where its marginal costs equal the price. The resulting pattern of abatement over activities, sectors, and countries is accomplished efficiently, at least cost. In principle, if regulators had perfect and detailed knowledge of all cost functions, command-and-control-type measures tailored to individual sources could be equally efficient. In practice, however, the information available to regulators is imperfect and incomplete. The parsimonious need for information favors using the market.

Third, there are literally millions of greenhouse gas emitters. They cannot be individually monitored. It is essential that each emitter respond to market and price incentives (and disincentives) to reduce their emissions. Prices do a better job than emissions police. However, some monitoring is still needed in tax and in cap-and-trade systems to

---

[10] This is an oversimplification. In fact, there are several greenhouse gases with different temperature-increasing potential and different atmospheric lifetimes. Conversion factors to establish their carbon equivalence must be calculated if there is to be a single product and price. Note that the efficient price is time specific.

[11] For example, carbon monoxide is more damaging when released at ground level in urban areas.

ensure that taxes are not circumvented and caps are honored. Subsidies for emissions reductions also require establishing BAU baselines in addition to monitoring.

Market-based systems provide a continuing incentive for polluters to reduce their emissions through technological change. In a tax scheme, the incentive is to reduce taxes by reducing emissions. In a cap-and-trade system, the incentive is to either reduce payments for permits or increase the number put up for sale. In contrast, in a regulatory approach once the standard has been met there is no incentive to reduce emissions further. And reducing abatement costs through technological improvement may simply lead to tighter standards. There is another potential advantage to some market-based systems. Both carbon taxes and auctioned tradable permits can generate revenue for the government.[12] In principle, this revenue can be used to reduce pre-existing taxes with distortive and inefficiency effects such as labor taxes. This is the so-called double-dividend hypothesis and is considered later. One final advantage is that carbon tax and cap-and-trade systems are directly targeted on the externality distortion. Often regulations are not finely targeted and create so-called by-product distortions. For example, fleet mileage standards for vehicles are only indirectly linked to carbon emissions and provide no incentive to reduce vehicle miles traveled. Indeed, by lowering driving cost, they do the opposite and emissions may increase.[13]

We should be aware, however, that a policy based on the pricing of carbon emissions works best when applied to the burning of fossil fuels. The amount of carbon released from various agricultural activities differs, depending on tilling practices, soils, timing, and other factors. In principle, the carbon emission consequences of different agricultural activities would need to be calculated if the activity itself were to be taxed (or subsidized). The carbon sequestration characteristics of forests also differ and need to be estimated before a carbon pricing policy is adopted. These are not simple tasks.

---

[12] Subsidies for emission reductions obviously do not generate revenues.

[13] Taxes are not immune to this defect. For example, a fossil fuel tax will provide no incentive to switch from high-carbon to low-carbon fuels (i.e., coal to natural gas). The need to closely target corrective measures to specific market failures is worked out in the trade literature as the theory of domestic distortions.

Emission Taxes versus Cap-and-Trade

Much has been written on this subject. In important respects they are similar, or can be made similar. Both put a price on carbon emissions and garner the important efficiency advantage of equalizing marginal costs among sources. Both create incentives for conservation and technological innovations. Both can be manipulated to modify their distributional impacts. Both require some system of monitoring and mechanisms for compliance. Both could be implemented at the national or international level, and if international coverage is incomplete, both can lead to "carbon leakage" (see Chapter 7). Also, if the permit market is competitive, the price of permits and the tax will be the same when the tools are used to obtain the same abatement objective.

Despite these similarities, there are differences. One obvious difference is that a carbon tax leads to a more certain abatement cost outcome but a less certain emissions outcome than would a cap-and-trade system.[14] If certainty of cost weighs more heavily than certainty of emissions level, a tax may be preferable, and vice versa if certainty of meeting an emissions target is more important. Pizer (2005), for example, emphasizes cost certainty and favors price measures over quantity measures and for much the same reason favors intensity targets over absolute targets. (The question of fixing prices or quantities has a long history in trade policy in the guise of tariffs versus quotas, with economists backing tariffs and the business community generally favoring the more certain quantity tool – quotas). The difference between price certainty and quantity certainty should not be exaggerated. Hybrid instruments consisting of a quantitative cap and carbon permit trading combined with a price ceiling and perhaps a price floor have been proposed. The price limits would be maintained by government's willingness to buy or sell permits as needed. Borrowing and banking provisions in a cap-and-trade system can also dampen price volatility. While reducing price uncertainty, these modifications obviously increase emissions uncertainty. Nordhaus has termed these

---

[14] Recall also from Chapter 5 that by fixing emission levels, a cap-and-trade system avoids the so-called green paradox.

schemes "prices in quantity clothing." Note, however, that a cap-and-trade scheme that allows regulated entities to cover their CO_2 obligations through offsets purchased from non-regulated emitters, perhaps developing countries, face additional uncertainty as the supply of those offsets is itself uncertain. A negative correlation between the supply and demand for offsets would increase price volatility (Fell et al. 2010).

The possibility of misestimating abatement costs may favor the tax tool.[15] If marginal abatement costs tend to be steep relative to marginal benefit curves, as they are likely to be in the case of global warming, an overestimation of marginal costs above their true level will lead to too much abatement under a tax and too little abatement under cap-and-trade. But the social costs of misestimating will be higher under the cap-and-trade. Underestimating abatement costs will also lead to incorrect abatement levels, and the social loss is again higher under the cap-and-trade than the tax tool. The reason marginal abatement costs are thought to be steeper than marginal benefits is that they are related to the *flow* of emissions, whereas in the case of global warming, marginal benefits (damages avoided) are related to the *stock* of greenhouse gases in the atmosphere. That stock accumulates relatively slowly. Therefore, the impact of a unit of abatement on the stock is small, and hence the marginal benefit of that unit is quite stable. This does *not* imply the marginal benefit of abatement is either "high" or "low." The argument loses force if the future holds unpredictable but major tipping points, in which case marginal benefits (damages avoided) can abruptly become very steep. It may also lose force if uncertain shocks are permanent rather than temporary.

Next, the two tools may differ in their fiscal impacts. This need not be the case. An efficient auction of permits under cap-and-trade should yield about the same revenue as an emission tax. But the political economy selling point for a cap-and-trade system is that at least some of the permits would be freely distributed rather than auctioned off. The most likely choice is a system of "grandfathering" permits to firms on the basis of historical emissions. The distributional impact of free allocation is important. In a tax system, carbon emitters are not only liable for their abatement costs, but must also pay the tax on their

---

[15] The classic paper is Weitzman (1974).

residual emissions. This can be thought of as a "rent" to society, the ultimate "owner" of atmospheric resources. In a cap-and-trade system, emitters as a group must pay for their abatement costs, but if permits are freely distributed, they do not pay any rent for residual emissions. In essence, they would be awarded a newly created, valuable property right to (limited) emissions. The prospect of at least some free permits explains why business is generally more receptive to cap-and-trade schemes than taxes.[16] This distributional distinction between taxes and non-auctioned permits does not, however, undermine the basic efficiency of a cap-and-trade system. A firm that is given permits will still consider the opportunity cost of using the permit and not abating. The opportunity cost is permit sales revenue forgone. That cost will be reflected in product prices and efficiency is promoted.

There is another consequence, associated with the so-called double-dividend question. In the 1990s, when economic tools were first considered seriously to deal with global warming, it was thought that the revenue raised could be used to offset other, distortionary taxes, such as taxes on labor and capital. The resulting efficiency gain would provide a double dividend – correcting the global warming externality *and* reducing inefficient taxes elsewhere in the economy. Subsequent research showed that things are not that simple. Increasing the price of carbon does address the environmental externality but will itself create a by-product distortion (called the tax interaction effect). The carbon tax increases prices generally, reduces real wages and labor supply, and distorts the labor market. Revenues raised by carbon taxes or auctioned permits could still be used to reduce distortionary capital and labor taxes, but the effect would be muted by the by-product distortion. If this later effect is sufficiently strong, the net tax interaction effect is negative.[17] Even so, auctioned permits or carbon taxes would be more efficient than freely distributed permits. The by-product distortion occurs under both arrangements, but only taxes and auctioned permits raise revenue and can finance reductions in (inefficient) taxes

[16] The two tools could have the same distributional impact if tax receipts were refunded to emitters in lump-sum fashion, independent of their abatement effort. This is unlikely. Grandfathering has the undesirable effect of discouraging new firms from entering.

[17] This suggests that the optimal (Pigouvian) carbon tax coming out of a general-equilibrium model that includes tax distortions may be lower than generally assumed.

on labor and capital. Parry (2003) shows that for the United States, the efficiency costs of auctioned permits or carbon taxes may be dramatically lower than in a grandfathered permit scheme. Indeed for moderate carbon taxes or auctioned permits, the efficiency costs may be negative – a welfare increase. But even this advantage can be questioned if the additional revenue is used unproductively by governments. There is little merit in substituting an efficient externality tax for an inefficient labor or capital tax if the proceeds are squandered. Finally, the tax interaction complication tends to partly rehabilitate command-and-control tools such as performance standards and technology standards. The reason is that these policies do not charge firms for residual emissions, reducing the negative tax interaction effect as compared to emissions taxes and auctioned permits.

Two other aspects of tax and permit schemes have drawn attention. First, within a few years, carbon taxes set at optimal levels (i.e., equal to marginal damages) could impose large burdens on energy-intensive industries. The conventional view is that this will generate massive opposition from sectors well known for their lobbying strength. Under these circumstances, it would appear that a cap-and-trade system can more easily "buy off" political opposition that would otherwise block a strong policy. This could be done by either exempting energy-intensive industries from the permit scheme or providing permit "rebates" to these firms. In fact, there is less here than meets the eye. A tax system that treated infra-marginal emissions as a property right, and only charged tax on marginal emissions, could accomplish the same results: efficiency in pricing marginal emissions, and a reduction of politically unacceptable tax payments (Pezzey 2003). Nor is it easy to distinguish the taxes from cap-and-trade schemes on the basis of encouraging non-productive rent-seeking activities. Free allocation of permits would certainly do so, but tax schemes are not immune from rent-seeking behavior either.

This leads to another point. If sector or country coverage by a carbon tax is incomplete, demand is shifted to the untaxed sectors and countries. Some mechanism for maintaining the output and employment of the taxed sectors may therefore appear desirable either to minimize carbon leakage or for equity reasons. As pointed out earlier, to relieve firms from infra-marginal taxes, output-based rebating have been suggested.[18] However, there is an efficiency cost. The rebates amount to an output subsidy. This introduces a bias between emission reduction via conservation (reduced energy use) and reducing the *rate* of emissions per unit output. The efficient balance between the two is thrown off. The bottom line is higher costs to achieve any given level of emissions reduction.

If the energy sector is to be coddled, it should not be overindulged. Carbon pricing increases the price of carbon-intensive products and, if full rebates to firms for carbon taxes were made, they may wind up with higher profits. Fischer and Fox (2009) have attempted to calculate the optimal tax and the optimal tax rebate for six energy-intensive sectors while taking into account two second-best distortions: the tax interaction effect describe earlier, and the prospect of carbon leakage through incomplete tax coverage. They use a computable general equilibrium (CGE) model to simulate the impact of a $50 per ton carbon tax. The results show that to maximize global welfare, full rebates of emission taxes to the electricity and the refined petroleum sector in the United States are not desirable (rebates should equal only 36 and 28 percent, respectively, when transportation is included in the regulations). On the other hand, however, the model suggests that the optimal rebates for the manufacturing sector exceed 100 percent. This surprising result apparently reflects the relatively low energy and carbon intensity of U.S. manufacturing, and thus the desirability of increasing U.S. share of world output.[19]

International Aspects of Taxes and Cap-and-Trade

To this point, we have assumed that tools would be deployed at the national level. We now offer a comment about the tools in an international context. Some governments may choose carbon taxes, some may

---

[18] Output-based rebates would distribute the carbon tax revenues back to firms in proportion to their share of the industry's output.

[19] In a study of U.S. fossil fuel industries, Bovenberg and Goulder (2001) found that to offset the negative impact on profits and equity values from carbon taxes would require rebates amounting to only a small fraction of their tax payments. In contrast, if firms' carbon tax payments were offset dollar for dollar by reductions in other taxes they pay, the firms would be substantially overcompensated. This is because the firms can pass along much of the carbon tax burden to consumers.

choose national-level cap-and-trade systems, and some may rely of technology mandates and performance standards[20] (indeed, some may stand back and do little or nothing). The difficulty that arises is not so much differences among countries in their choice of tools, but inconsistent targets. If one country is strict and another country is not, the price of carbon would vary, the efficiency gains of equalizing carbon price would be lost, and undesirable carbon leakage and shifts of competitive advantage would result. This would lead to uncoordinated targets in either a tax, a cap-and-trade, or a mixed system. A comprehensive international agreement on either an internationally harmonized tax or an international cap with permits allocated by country is desirable. Within that agreement, countries could choose their preferred tools.

There is another wrinkle. As explained in Chapter 9, it will generally be necessary to have international financial transfers to obtain the widest possible membership in an international climate agreement. Presumably the transfers would flow from North to South. If all countries use a tax scheme or if all countries use a cap-and-trade scheme, transfers to induce participation are technically possible. In a cap-and-trade scheme, the financial transfers are embedded in the international distribution of emissions permits. The transfers are then activated with private-sector purchases of emissions permits. In a tax scheme, however, they presumably would be explicit government-to-government payments, which seems unlikely in the extreme. The alternative incentive to participate would be differentiated tax rates, which would be inefficient. This appears to favor cap-and-trade as providing a partial veil over the transfers, although money would subsequently cross borders and change hands when permits are traded. (CDM payments across borders seem to have been accepted, but the amounts remain relatively modest). In any event, the South could generate the emission reduction credits for sale through any means it wishes, including taxes and technology mandates.

Subsidies: The Other Market-Incentive Tool

Government subsidies flow through three spigots: one is unattractive; the second borderline; and the third attractive if carefully

---

[20] Technology standards in the energy sector are important in countries where governments, not the market, control investments.

administered. The first spigot is subsidies for fossil-fuel-based energy production and consumption, which remain widespread, especially in developing countries. Preliminary estimates of these subsidies are on the order of $600 billion (1.2 percent of world GDP).[21] These subsidies increase energy consumption and carbon emissions; they discourage the use of renewable energy sources; and their financing draws resources away from more productive uses. This spigot should be turned off.

The second spigot has more promise. It links a subsidy to achieving emissions reduction. The government specifies a baseline emissions level and pays for abatement below that level. Instead of being beaten with the tax stick, the firm is offered a subsidy carrot. Emissions would then carry an opportunity cost – subsidy revenues forgone – and emissions would have an implicit price. Abatement will presumably be carried out to the point where marginal abatement costs are equal to this price – the same result as an equivalent tax. Firms are delighted but taxpayers less so. Economists would be pleased with the efficient abatement by the firm and by the equalization of marginal abatement costs among firms in the sector. However, they would be cautious about three features. First, the subsidy presents an invitation to firms to manipulate their baseline emissions upward so they can receive a larger subsidy. The cost would be twofold: distorted emissions in the base period and extortion of the taxpayer. Second, these subsidies discourage firms from exiting the industry. Whereas emissions from individual firms will fall, the level of industry emissions will be higher that desirable. Third, we have seen that carbon tax revenues can be used to reduce other distortive taxes – the double dividend. Emissions reduction subsidies earn nothing but instead need additional revenue from these inefficient sources.[22] This can be thought of as a negative dividend.

The third subsidy spigot addresses failures in the innovation "market." It does so by subsidizing R&D in renewable energy and carbon capture and storage technologies, and by subsidizing output

---

[21] Burniaux et al. (2008) citing preliminary estimates by the Global Subsidy Initiative.

[22] The Clean Development Mechanism has some of the characteristics of an emissions reduction subsidy and is vulnerable to baseline manipulation. One could argue that differences in the marginal utility of income between rich and poor countries might compensate for inefficiencies inherent in raising revenue to fund the subsidy.

expansion of renewable to capture learning-by-doing cost savings. The rationale is straightforward. Pricing carbon emission responds directly to negative greenhouse gas externalities, and indirectly through a change in relative prices, to encouraging renewable energy sources. It does not, however, directly address the positive externalities from innovation. Another, supplemental, policy is needed. There is symmetry at work here. The negative incentive of taxes is directed toward the negative externality, and the positive incentive of subsidies is directed toward positive innovation externalities. However, as with most government interventions, the devil is in the details. Not all renewable energy technologies are equally deserving of subsidies. Many have undesirable side effects. Some, such as ethanol, represent a mature technology but continue to enjoy large subsidies.[23] Fischer and Fox (2009) find that R&D subsidies for renewables are particularly inefficient as they postpone the replacement of fossil fuels until after costs are brought down. Breakthrough technologies such as carbon capture and storage might tell a different story. Careful selection and cost-effective administration of this category of subsidies is the challenge.

Summary

Several points stand out. First, analysis of policy tools has made substantial, if incremental, progress. The prospect of global warming has brought economics to a higher level in evaluating policy tools under uncertainty, in understanding interactions in second-best, distorted systems, and in illustrating trade-offs among competing evaluation criteria. Second, there remains a remarkable level of agreement among economists that pricing greenhouse gas emissions and especially carbon must be the centerpiece of a serious climate policy. Third, perceptions may play an important role in choosing targets and tools. Emissions intensity targets are less transparent than absolute quantity targets, and may be easier to swallow. However, this blurring is a potential source of mischief. Along the same lines, an international cap-and-trade system bundles distributional and efficiency effects into permit allocations and is less transparent than a uniform tax with negotiated

---

[23] Burniaux et al. (2008) report that the implicit costs of ethanol subsidies typically exceed $300 per ton of CO_2 avoided – far above alternative abatement opportunities.

transfers among countries. This can make cap-and-trade more accept-able, but that result is not assured and remains speculative. Fourth, in general, subsidies should be kept in the technology toolbox, not the mitigation toolbox.

References

Bovenberg, L. and L. Goulder (2001). Neutralizing the Adverse Industry Impacts of CO_2 Abatement Policies: What Does It Cost? In *Behavioral and Distributional Effects of Environmental Policy*, C. Carraro and G. Metcalf (eds.). Chicago: University of Chicago Press.

Burniaux, J., J. Chateau, R. Duval, and S. Jamet (2008). The Economics of Climate Change Mitigation: Policies and Options for the Future. *OECD Economics Department Working Paper 658*.

Ellerman, D. and S. Wing (2003). Absolute vs. Intensity-based Emissions Caps. *MIT Joint Program on the Science and Policy of Global Change. Report No. 100.*

Fell, H., D. Burtraw, R. Morgenstern, and K. Palmer (2010). Climate Policy Design with Correlated Uncertainties in Offset Supply and Abatement Costs. *Resources for the Future Discussion Paper 10–01.*

Fischer, C. and A. Fox (2009). Combining Rebates with Carbon Taxes: Optimal Strategies for Coping with Environmental Leakage and Tax Interactions. *Resources for the Future Discussion Paper 09–12.*

Fischer, C. and R. Newell (2008). Environmental and Technology Policies for Climate Mitigation. *Journal of Environmental Economics and Management* 55: 142–62.

Goulder, L. and I. Parry (2008). Instrument Choice in Environmental Policy. *Review of Environmental Economics and Policy* 2 (2): 152–74.

Karp, L. and J. Zhao (2009). Suggestions for the Road to Copenhagen. Report of the Experts Group on Environmental Studies. Swedish Ministry of Finance, Stockholm.

Parry, I. (2003). Fiscal Interactions and the Case for Carbon Taxes over Grandfathered Carbon Permits. *Oxford Review of Economic Policy* 19 (3): 385–99.

Pezzey, J. (2003). Emissions Taxes and Tradable Permits: A Comparison of Views on Long-run Efficiency. *Environmental and Resource Economics* 26: 329–42.

Pizer, W. 2005. Climate Change Policy under Uncertainty. *Resources for the Future Discussion Paper 05–44.*

Schmidt, J., N. Helme, J. Lee, and M. Houdashelt (2008). Sector-based Approach to the Post-2012 Climate Change Policy Architecture. *Climate Policy* 8: 494–515.

Timilsina, G. (2008). Atmospheric Stabilization of CO_2 Emissions: Near-term Reductions and Absolute versus Intensity-based Targets. *Energy Policy* 36: 1927–36.

Weitzman, M. (1974). Prices vs. Quantities. *Review of Economic Studies* 41: 477–91.

7

Trade and Global Warming

This chapter examines the points at which climate and climate policy intersect with the international trade system. The issue of trade and environment dates to the 1972 United Nations Conference on the Human Environment (Stockholm Conference), and the publication in the same year by the OECD of its "Guiding Principles" concerning the international economic aspects of environmental policies.[1] At that time, the principal concerns were the effects of environmental regulations on trade: the competitive impact of differences among countries in pollution abatement stringency; the use of environmentally related product standards as covert trade barriers; and the appropriate use of trade measures to induce or coerce trade partners into altering their environmental practices. Subsequently, an additional concern was added – the effects of international trade on the environment, and more specifically for us, the effects of trade on global warming. All of these concerns are very much with us today in the debate on trade and global warming policy.

This chapter starts with the more recent issue first – the effect of trade and trade liberalization on climate change. It then takes up the important and controversial issue of the impact of global warming policies on international competitiveness, and the related carbon leakage question. The following section examines carbon labeling, "food miles," and related issues. The final sections consider permit trading

[1] The most significant was the polluter pays principle (PPP), which has had a curious history. It was originally proposed by the OECD to rule out government subsidies for pollution abatement in the private sector to prevent trade distortions. Since then, it has taken on an ethical dimension. See Pearson (2000).

and the so-called Dutch disease, as well as the manipulation of international trade in emissions permits.

The Impact of Trade on Global Warming

Analytical Approaches

Is international trade good or bad for the environment? Does trade liberalization contribute to global warming? Three approaches have been used to answer these questions. The dominant approach traces back to an *ex ante* study of the North American Free Trade Agreement (NAFTA) by Grossman and Krueger (1993) and later refined by Copeland and Taylor (2003). In this framework, the effects of trade liberalization are transmitted through three channels: a scale effect, a composition effect, and a technique effect. The scale effect measures the increase in output and pollution associated with trade liberalization and thus has a negative impact on the environment.[2] The composition effect measures changes in the structure of output as it shifts in response to trade opportunities. The effect of liberalization will reduce pollution and have a positive impact on the environment if the economy has a comparative advantage in "clean products," but will increase pollution and degrade the environment if it has a comparative advantage in "dirty" products. (This terminology, although widely used, is not very accurate. It is not the product but the production process that is considered polluting). The technique effect measures the response of environmental policy to increases in real income associated with trade liberalization. If policy responds, pollution abatement standards are presumably tightened, and firms respond by using cleaner production techniques with a positive impact on the environment. The net impact of these three forces then depends on the size of the scale

---

[2] Note the assumption that pollution is related to the level of output. If pollution were linked to the level of *input*, the scale effect would not be reliable because trade does not directly affect inputs available to the economy. For example, the liberalization of agricultural trade may leave cultivated land unchanged but increase output by relocating crop production to the most suitable soils and climates (Ricardo's Portugal-wine example). There would be no reason to expect that input-linked environmental degradation such as soil erosion would increase. Note also that this tripartite approach pays little attention to pollution arising from consumption, the level and composition of which is also affected by trade.

effect (negative), the size of the technique effect (positive), and the comparative advantage of the economy in question in clean or dirty products.[3] An important conclusion is that if efficient environmental policies are in place, trade is welfare-improving. However, if environmental policies are too weak or are absent, the benefits of trade can be outweighed by losses from increased environmental degradation. Thus improvements in environmental policy may be required.

This framework is useful in sorting out the interactions of trade and environment, explaining the so-called Environmental Kuznets Curve (EKC), and investigating the pollution haven hypothesis, but is not very helpful in the case of global warming. The EKC, first proposed by Grossman and Krueger, has an inverted-U shape, with pollution first rising with per-capita income, reaching a maximum, and then declining at higher income levels.[4] The potentially helpful role of trade is in speeding countries to the declining portion of the curve. In contrast, the pollution haven hypothesis suggests that countries may attempt to attract polluting industries on the basis of environmental policies that are deliberately weak as compared to trade partners.[5]

Although we are interested in these issues, and particularly the pollution haven question, the tripartite classification of scale, composition, and technique effects were designed to study environmental impacts from local, not transnational or global pollution. We are interested in the *global* level of greenhouse-gas emissions. Shifts in the structure of production in one country (the composition effect) may be fully offset by opposite shifts in other countries, with little change in global emissions. Moreover, the technique effect – the inducement to tighten pollution standards as income increases – is apt to work very

---

[3] Dean (2002) attacks this problem using a simultaneous equation system in which the "supply" of the environment is modeled as an endogenous factor input to production. Applying the model to China, she finds evidence that trade-induced changes in relative prices do indeed lead to an increase in pollution, but the increase in income due to trade works in the opposite direction – a positive "technique" effect on the environment.

[4] It was named after Simon Kuznets, who found a similar inverted-U shape between income inequality and per-capita income. The inverted U was based on cross-country data at a point in time and does not indicate the path that all countries must take.

[5] The evidence for the pollution haven hypothesis is mixed, but not strong. For example, Dean et al. (2009) find support when looking at equity joint ventures in China when the source of funding is through Hong Kong, Macao, and Taiwan, but no evidence when investments are funded from other sources, including the West.

differently with transnational pollution such as carbon emissions, if it works at all. As explained in the next chapter, emission restraint by one country confers benefits on all others. The incentive to establish stronger abatement targets as income increases is weakened when the benefits go abroad.

The second approach to examining the effect of trade on global warming is more suitable to the transnational character of the problem. Copeland and Taylor (1995) construct a model in which two groups of countries (high- and low-income North and South) follow endogenously determined environmental policies that maximize their narrow national welfare. Pollution is a global public "bad." In this world, moving from autarky to trade does *not* increase pollution, although its source is shifted from North to South, provided that the income differences are not too large. The essential reason is that the environment is treated as a factor of production (pollution is considered an "input" into production), and, according to the well-known factor price equalization theorem, trade leads to the international equalization of factor and hence of pollution permit prices. As factor prices equalize, the relative use of "pollution" and other inputs to production equalize in the North and South. Whether pollution-intensive products are sourced in rich or poor countries, the technique of production is the same. Free trade in permits would have the same effect – either free trade in goods or free trade in permits equalizes the price of pollution and prevents increases in global pollution via pollution havens. To obtain this encouraging result, however, the North must tighten its environmental standards as the source of pollution shifts to the South. The typical country in the North faces more foreign pollution, which increases the marginal damage from its domestic emissions, hence justifying a more stringent policy.[6]

The third analytical approach rejects the assumption that all countries follow optimal environmental policy (Chichilnisky 1994). Instead it emphasizes that environmental resources are likely to exist as open-access, common-property resources, especially in poor countries. With inadequate property rights, the South makes excessive use of these

---

[6] There is a whiff of extortion in the model. The South, with its lower income, is able to credibly commit to greater pollution with the opening of trade. However, the model does not consider differential damage functions as between North and South, which may affect such strategic behavior.

resources and sells its exports at less than social cost. International trade is distorted, and the North contributes to the problem by over-consuming the underpriced, resource-based exports of the South. Whereas trade may be the proximate cause of environmental degradation, the ultimate cause is a failure to properly price environmental inputs to production. Although this theory was developed in the context of natural resources such as forests, grazing lands, and fisheries, it sheds some light on excessive use of the atmosphere as a sink for CO_2 emissions. It also reminds us that partial restrictions on abusive use of resources by some countries may simply drive the activities to less regulated sites. This, however, is not directly relevant to climate change. To adapt it to global warming, one would need to acknowledge the transnational character of the damages and recognize that property rights to the atmosphere remain inadequate in the North as well as the South.

Additional Connections Linking Trade to Global Warming

Two other connections, one positive and one negative, link trade to emissions and have received considerable attention. First, trade liberalization in climate-friendly goods, services, and technology clearly has a positive impact on emissions-mitigation efforts by reducing price and increasing access to mitigation technologies. A recent joint report by WTO and UNEP (2009) reviews these efforts and provides examples, data, and sources. The flow is not always North to South. The report notes that a number of developing countries have significant exports of products within the renewable energy products category. China's success with solar technologies and wind turbines is often mentioned in this regard.

The other connection is far more controversial. Trade drives a wedge between where a good is made and where it is consumed. That means transportation, and transportation means emissions.[7] The effect of trade and trade liberalization on emissions from transportation is embedded in a larger debate on measuring "carbon footprints" and carbon labeling (e.g., "food miles"), which we consider later in the chapter. At this point, we wish to stay with the narrower

---

[7] Some trade in services requires transportation (e.g., tourism), but some does not (electronic banking). Domestic trade also requires transportation.

question of whether trade liberalization increases carbon emissions from transportation. The evidence suggests that it does, and by substantial amounts.

The question involves a number of linkages: from trade liberalization to changes in the commodity composition and geographic pattern of international trade; the resulting changes in the use of alternative transportation modes (summarized as land, maritime, and air); and estimates of the carbon emissions per ton kilometer of various transportation modes. The end point is estimates of changes in greenhouse-gas emissions. Hummels (2009) has provided preliminary estimates. The analysis proceeds in steps. A trade model is used to estimate the value of trade changes, by country and product, if there were full trade liberalization; the values are converted to quantities; with data on transportation mode use by commodity and changes in distances and increases in quantities traded, the increase in ton-kilometers by mode of transportation is estimated; and using data on CO_2 emissions per ton-kilometer by transportation mode, the total increase in emissions is calculated. A more ambitious but difficult model would then feed modal demand changes, through modal prices, back to international trade estimates.

The results show that with full trade liberalization, the value of trade would rise by 5.8 percent, and the quantity of emissions would rise by up to 10 percent. This result reflects the commodity composition of increased trade (indirectly mirroring the current pattern of protectionism), changes in the geographic pattern of trade, and the great disparity in emissions per ton-kilometer as between maritime, land, and air transport.[8] In particular, full trade liberalization would eliminate the preferential treatment of partners in regional trade arrangements in Europe, North America, and Asia. This leads to greater long-distance trade, a contraction of road and rail transport, and an expansion of the least and most carbon-intensive transportation modes – by sea and by air.[9] Recall also that these results pertain

---

[8] Measured in grams CO_2 per ton-kilometer maritime transport ranges from 4.5 (dry bulk) to 16.3 (LNG), land transport from 22.7 (rail) to 119.7 (road), and air from 552 to 1,020.

[9] The numbers can be quite interesting. Currently for bulk agriculture, 96.8 percent of ton-kilometers are by sea, 0.4 percent by air, 1.3 percent by road, and 1.6 percent by rail. With full tariff liberalization, the percent increases in bulk agriculture ton-kilometers are 71.8 for sea, 18.8 for air, 7.8 for road, and 15.3 for rail. For textiles, currently

to trade liberalization only. The normal growth of trade will dwarf the impact of liberalization.

The Impact of Global Warming and Global Warming Policy on Trade

We now shift from the effects of trade and trade policy on global warming to its converse, the effects of global warming and of global warming policy on trade. The effect of global warming on trade is, by itself, not a very interesting question. Chapter 4 identifies some of the environmental and economic damages that might be expected from warming, which is a more important question. It is sufficient to say that there will be trade effects. They will be most pronounced in natural resource sectors and especially agriculture, and in natural-resource-based economies; the international trade system will help moderate severe local effects, for example from localized drought; trade will also spread the damages costs internationally via terms of trade changes.

The interactions of global warming *policy* and the trade system are more interesting. There are essentially three issues: carbon leakage and competitiveness, climate-related non-tariff trade barriers, and the use of trade carrots and sticks to promote climate cooperation.

Carbon Leakage: Concepts

The conventional view is that if one group of countries – for example, Annex 1 countries under the Kyoto Protocol – set out to restrict carbon emissions among themselves through economic instruments, they would encourage increased emissions by non-restrained countries. It is worth noting that one can introduce trade into the analysis in a way that challenges the inevitability of this result. Copeland and Taylor (2005) construct a static North–South model consisting of a "clean" and a "dirty" good, with welfare-maximizing, endogenous environmental policy (governments using auctioned permits) in both regions. Emissions lower utility but do not affect production functions. The starting point is a non-cooperative Nash equilibrium in which each government chooses an emissions target to maximize its welfare,

76.8 percent of ton-kilometers are by sea, 6.8 percent by air, 0.1 percent by rail, and 16.3 percent by road. With liberalization, the percent increases in textiles are 37.8 by sea, 26.7 by air, −24.4 by rail, and −38.2 by road.

treating the rest of the world's emissions as fixed. If in autarky North reduces its emissions, South responds by increasing its emissions. This is the strategic, free-rider effect. If instead of autarky, trade is introduced, the response is more complicated. In addition to the free-rider effect, there is a substitution effect and an income effect in the South. If the South is a dirty-good exporter, the terms of trade move in its favor and output and emissions increase further. The income increase acts in the opposite direction, however, leading the government to tighten emission standards. The latter effect may dominate. If so, the conventional view needs modification. Emissions reduction in the North may lead to emissions reductions in the South, and the carbon leakage problem evaporates.[10]

Before celebrating this happy day, one might wish to consider whether a static model captures the essence of the problem. The strategic free-rider response, which is present in both autarky and trade, implies that the South anticipates a real income gain in the future, perhaps generations from now, and responds by allowing emissions to rise today. The inter-temporal optimization is inherently dynamic. In contrast, the terms of trade income change is current. It is unclear whether simply summing the emissions response is adequate. At a more basic level, one can question whether the assumption of welfare-maximizing endogenous emissions policy describes climate change reality.

Returning to the more conventional view, three channels for carbon leakage are cited. First, the Annex 1 reductions would weaken the incentive for reduction elsewhere as marginal damages are reduced. Reductions in Europe under the Kyoto Protocol reduce the incentives for others, including the United States, to limit emissions. This is the strategic free-rider effect. This channel is independent of trade. Second, non-restrained suppliers gain a competitive advantage and increase production of emission-intensive goods, as they do not have to pay for emission permits or taxes.[11] The environmental cost of this measured in emissions would be compounded if the unrestrained suppliers were inefficient energy users. Third, the emission restrictions would reduce the demand for and price of fossil fuels, thus increasing their

---

[10] Endogenous technology can also confound the carbon leakage proposition. See Glombek and Hoel (2004).

[11] Some commentators assert that multinational corporations would be responsible for the shift, although there is no evidence that they would take the lead.

use in unrestricted countries.[12] These shifts are the heart of the carbon leakage problem and are of central importance. They underscore the desirability of a universal emissions control system – either a harmonized tax or global cap-and-trade. They highlight how environmental objectives can be compromised by non-participation. They are at the core of political economy opposition to a strong climate policy that would, allegedly, weaken the competitive position of energy-intensive sectors. And they are used to justify questionable policies that would reduce leakage.

The argument is plausible and is familiar to anyone who has worked in the murky world of the second-best. Partial removal of an economic distortion – whether a trade restriction or one resulting from an externality such as carbon emissions – is no guarantee that welfare improves. The empirical evidence has to be examined. In the case of preferential trade, trade creation must be balanced against trade diversion. Something similar is suggested with carbon leakage. Incomplete control of emissions compromises the climate objective and increases the cost.[13] A welfare improvement is no longer automatic.

Carbon Leakage: Estimates and Policy Responses

The severity of carbon leakage – how much it undermines emissions targets, how much it contributes to higher abatement costs, and how potent it is in mobilizing opposition to emissions reductions – awaits more evidence from a serious abatement effort. Much will depend on the carbon price differential between restricted and non-restricted countries (i.e., the stringency of targets), the number and size of countries restricting emissions, the trade elasticities of emission-intensive industries, and investment location variables.

Ex post studies are rare because emission-reduction efforts have been rare. Barker et al. (2007) have looked for evidence of carbon leakage as a result of environmental tax reforms in six EU countries during the period between 1995 and 2005. The size of the tax revenues was quite modest, ranging from .07 percent to 1.08 percent GDP. The results show very little, if any, carbon leakage. In another study,

---

[12] The stronger the second effect, the weaker the third effect.

[13] A partial ban on deforestation has the same effect – pressure is shifted to uncontrolled forests.

Reinaud (2008) concludes that there is no evidence of carbon leakage in heavy industries in the first two-year phase of the EU's Emissions Trading System. She attributes this to generous free allocation of allowances, the general boom in prices, and the very short time period. She cautions that this may change as emissions restrictions tighten.

Ex ante estimates are accumulating but show considerable variation. At the high end, Babiker (2005) assumed an oligopolistic market structure for energy-intensive industry, increasing returns to scale, and strategic behavior. He found leakage rates of up to 130 percent, implying a net *rise* rather than a reduction in global emissions, but this is considered by many to be too high. In its 2001 report, the IPCC gives an estimated leakage rate of 5–20 percent. Its 2007 report provides a brief review of the more recent studies, mostly in the same range, but sometimes higher in narrowly specified industries.

As the United States moves closer to a formal mitigation policy, competitiveness and leakage studies are being prepared. One approach is to examine output responses by industry to energy price changes over the past two decades, and to infer from those statistics the effects of a carbon-pricing scheme. Aldy and Pizer (2009) follow this approach, using a price of $15 per ton CO_2 for illustration. They find that significant production declines (in the range of 3–4 percent) would be limited to certain emission-intensive industries, for example segments of the iron and steel industry, primary aluminum, cement, paper, glass, and fertilizers.[14] The majority of the decline is attributed to reduced domestic consumption due to price increase rather than increased foreign competition (carbon leakage). This suggests that full border adjustments may not be needed.

Ho and his colleagues (2008) have also investigated the competitiveness and carbon leakage effects of pricing carbon in the United States, in this case at $10 per ton CO_2. They find relatively high leakage rates. Twenty-five percent of U.S. reductions in emissions are offset by induced increases in foreign emissions working through trade changes, and this rises to more than 40 percent for certain energy-intensive sectors. One helpful feature of this study is that it examines the time path of adjustment, an important element in policy design.

---

[14] An average production decline of 1.3 percent across all industries, of which about half is due to carbon leakage.

The most general conclusion is that the short-run output declines in sensitive industries shrink over time as firms adjust their inputs and employ new technology.

In response to pending legislation creating an economy-wide cap-and-trade system (HR 2454), an interagency report (U.S. Government 2009) analyzed competitiveness and leakage concerns for U.S. energy- and trade-intensive industries. The base projected price was $20 per ton CO_2e. The modeling results showed that the inclusion of two provisions to shield energy- and trade-intensive sectors – permit allocations to local electricity distribution companies and output-based allocations to energy- and trade-intensive sectors – would virtually eliminate carbon leakage and trade-competitive losses.

Firms are understandably concerned with the effect of emission controls on profits. Goulder et al. (2009) uses a numerical general equilibrium model to focus on profits by industry under various emission allowances allocations. They find that free allocation of only about 15 percent of total allocations would compensate for profit losses in the most vulnerable industries. One-hundred-percent free allocation would substantially overcompensate industry and would dissipate revenues that could allow reductions in other, distortionary, taxes (i.e., squander the double dividend).

Border Tax Adjustments

Chapter 6 and the immediately preceding study conclude that negative trade-competitive effects could be moderated with either carbon tax rebates or emissions allocation rebates to energy-intensive industries, but there would be an efficiency cost. Border adjustments – carbon tariffs on imports and carbon tax rebates to exports – have been discussed as an alternative. They, too, have efficiency costs but can be seen as having coercive value, nudging competitors toward greater emission reductions. For expositional reasons, we concentrate on competitiveness and defer a discussion of using trade measures to induce participation in a climate agreement or to enforce its obligations until the following chapter.

Competitiveness concerns arising from international differences in environmental policy have a long history in the United States (Pearson 2004). For example, Section 6 of the 1972 Federal Water Pollution Control Act required annual studies of the trade impact

of the legislation. Border tax adjustments consisting of import sur-
charges and export rebates to equalize environmental control costs
internationally were featured in the 1977 Copper Environmental
Equalization Act (not enacted), and have been proposed from time
to time ever since. They have been opposed on economic, legal, and
pragmatic grounds. The objections remain valid for local environmental problems, but there are several reasons for revisiting them in the
context of global warming and carbon leakage.

The original opposition to border adjustments was based on three
arguments: given international differences in environmental assimila-
tive capacity, income, and preferences, there is no economic efficiency
reason for equalizing environmental control costs internationally;
sovereignty allows countries to choose their own environmental
standards; in the absence of any international agreement, leaving the
determination of border adjustments to domestic authorities would be
an invitation for covert protection and would leave the international
trade system in tatters. The first reason is not strictly true in the case of
carbon. Even though there is still no efficiency reason to equalize total
mitigation costs internationally, efficiency does require equalizing the
price of carbon internationally. The second argument also loses force.
Global warming is a transnational problem involving international
externalities and not a purely domestic matter. Finally, the arbitrary
and potentially protectionist nature of border adjustments may be limited *if* the adjustment is based on a widely adopted climate agreement
that yields a specific price per ton of carbon emissions. This last point
is arguable and has not yet been demonstrated. Calculating embodied
emissions product by product and country by country would be tricky
and could be a source for mischief. Fluctuations in carbon prices and
complicated supply chains would pose other challenges.[15]

Setting aside the question of whether border adjustments could
enlarge a coalition committed to emission reductions, what would
be their environmental and efficiency consequences? To answer this
question, it is useful to distinguish among three types of border adjustments. The first is a tariff on imported goods set equal to the carbon

---

[15] Presumably the carbon tariff on imported goods would depend on the differential in
carbon prices between the importing and exporting country as well as the direct and
indirect carbon content of the imported good.

tax paid by domestic firms producing like products. This would be calculated by multiplying the domestic and foreign carbon price differential by the direct and indirect carbon content of the *domestic product*. If the foreign country, had no carbon policy only the domestic carbon price is needed. The second scheme is a tariff on imported goods set equal to the foreign and domestic carbon price differential multiplied by the direct and indirect carbon content of the *imported product*. The third scheme would add to the second scheme a rebate for exports, with the amount of the rebate equal to the direct and indirect carbon taxes paid by exporting firm on its domestic sales of like products.[16] (Export rebates of taxes on inputs to production fall into the long-standing and arcane legal issue of the treatment of *taxes occultes*).

A few comments are appropriate. As compared to no border adjustment, the first scheme would tend to shift production to the home economy and reduce carbon leakage. Global emissions would tend to fall if production in the home country is less carbon-intensive than in the country from which the imports originate. This is most likely to be the case in U.S. trade with many developing countries. Mattoo et al. (2009b) report that the carbon intensity of manufacturing in China is more than four times the intensity in the United States.[17] As compared to the first scheme, the second scheme will have a stronger effect on reducing emissions, again assuming the home country is less carbon-intensive. But the size of the tariff would be greater and the disruption to trade greater. Both the first and second schemes would increase domestic price and discourage emissions from domestic consumption. Notice also that under both the first and second schemes, foreign countries would gain no "credit" for attacking global warming through policies other than pricing carbon – for example, by subsidizing nuclear, wind, and solar energy. Despite these efforts they would be confronted by carbon tariffs.

---

[16] A numerical example might help keep this straight. Suppose the price of carbon is $30 in Home country and $10 in Foreign country, and that direct and indirect carbon embodied in a ton of steel produced in Home is 5 tons and is 12 tons in Foreign. The tariff on a ton of imported steel would be (30–10) × 5, or $100 in the first scheme and (30–10) × 12 or $240 in the second scheme. The third scheme would continue a $100 tariff on imported steel but would give a rebate on exports of $100.

[17] Direct and indirect carbon emissions per dollar of manufacturing output.

The final comment deals with the third scheme, which includes an export rebate. A long-standing proposition in international trade is the Lerner Symmetry Theorem, which states that with balanced trade a tax on imports is in effect a tax on exports. It follows that unless carbon taxes on exports are rebated, they are subject to double taxation if either scheme one or two are implemented. To regain tax neutrality (and efficiency) between import-competing and export sectors, scheme three proposes carbon-based import tariffs and carbon-based export rebates. This is likely to further reduce competitive concerns in the United States but may weaken global emissions reductions. This third scheme has an awkward appearance – rebating a tax on a global externality sounds perverse, and may have to be explained and justified on political economy, not environmental, grounds.

Mattoo and his colleagues (2009b) have attempted to quantify some of these results. They employ a multi-sector multi-country CGE model to estimate production and trade consequences of the three border adjustment schemes described earlier. The analysis assumes high-income countries take on carbon emission reduction cuts of 17 percent below 2005 levels by 2020. Low- and middle-income countries are assumed to take no action. The analysis includes border adjustments applied to all merchandise trade, and a separate analysis of only adjusting energy-intensive merchandise imports.

The good news is that even in the absence of border adjustments, the amount of leakage appears small. The reduction in high-income countries results in a 1 percent increase in emissions in low- and middle-income countries (in a technical appendix, the authors struggle to explain why the leakage rate is well below other estimate). A competitiveness concern remains, however. In the absence of border adjustments, U.S. production of energy-intensive manufactures declines by 3.5 percent and exports of these goods decline by 11.6 percent. As expected, all three schemes would reduce the decline in U.S. energy-intensive manufacturing and reduce imports of these items. Scheme three, which includes export rebates, results in an increase in U.S. exports of these goods.

The most interesting result, however, is the greatly different impact on the trade of developing countries of using domestic carbon content (scheme one) or foreign carbon content (scheme two) when calculating the import tariff. Under scheme one, China would face a carbon tariff of 3.1 percent on all manufacturing and 6.2 percent on energy-intensive manufactures when selling in high-income markets. The corresponding numbers for India are 3.5 and 6.8 percent. Under scheme two, China would face a carbon tariff of 26.1 percent on all manufactures and 42.7 percent on energy-intensive manufactures. The corresponding numbers for India are 20.3 and 28.5 percent. Under scheme one, China's total manufactures exports would fall by 3.4 percent and India's by 3.2 percent. Under scheme two China's total manufactures exports would fall by 20.8 percent and India's by 16.0 percent. These remarkable numbers are a direct reflection of the high carbon intensity of manufacturing in these two countries as compared to high-income countries. They pretty well scotch the idea of using foreign carbon content (scheme two) in setting import tariffs. One could argue that steep tariffs might induce a serious carbon-pricing policy in exporting countries, and therefore need not be applied. But in a globalized economy with large financial flows and tensions over exchange rates, such aggressive tariff tactics can backfire.

Border Adjustments: Legal Issues

There is a disconnect between trade law as embodied in the General Agreement on Trade and Tariffs (GATT) and the WTO and the economics of trade. GATT and the WTO appear to champion mercantilism, with tariff reductions termed "concessions" and countries requiring "compensation" for loss of export markets but not for loss of import access. This disconnect carries over to the treatment of indirect taxes such as value-added taxes. GATT allows such taxes to be applied to imports at the border and rebated on exports. The intent clearly is to protect producer rather than consumer interests. The same mindset supports carbon-related border adjustments. Whether these adjustments would be considered compatible with GATT/WTO obligations has been extensively discussed but remains unknown.[18] Rather than rehash what has already been written elsewhere, let us make three points. First, the use of trade measures, including border adjustments,

---

[18] For a good review see, the UNEP-WTO joint report, Trade and Climate Change (2009). An important slice of the debate is whether WTO allows countries to restrict imports based on the environmental manner in which the products were made or harvested. This is the "process and production methods" issue. See also Hufbauer, Charnovitz, and Kim (2009).

for ostensibly environmental purposes is more likely today than it was twenty years ago. Both public attitudes and recent WTO rulings contribute to this shift. Second, the specifics may be important – whether a carbon tax or a tradable permit scheme (a regulatory measure) is used, how permits are allocated, and whether border adjustments are part of a multilateral agreement or are unilaterally established. Third, unlike the Montreal Protocol, which also contains trade provisions but was never challenged, carbon-related border adjustments will surely be contested. It would be unfortunate if response to carbon leakage led to an unraveling of a somewhat fragile multilateral trade system.

Food Miles, Carbon Labeling, and Other Trade Issues

Competitiveness issues have dominated the environmental policy–trade discussions, but there has been a persistent, if more muted, interest in how policies that ostensibly protect the environment can become non-tariff trade barriers (NTBs). Governments have shown great ingenuity in designing product regulations in a fashion that penalizes imports and protects domestic producers. The great burst of environmentally related product regulations and standards that started forty years ago expanded these opportunities. Some were legitimate efforts to protect health and safety and domestic environmental resources, and some had questionable protectionist motivations. Aircraft noise standards, pesticide residue standards in food products, and turtle exclusion devises in shrimp fishing have all been subject to trade disputes. An even more specialized source of friction has been the introduction of voluntary (and occasionally mandatory) eco-labels. The question has re-emerged under the general rubric of carbon labeling and the controversy over "food miles." It may be worthwhile to sort through the issues.

Accurate and unbiased eco-labeling, including carbon labeling, has the potential to improve market efficiency. Labels help match preferences to the market. They also reduce the need for clumsy government regulation, as was evident in the notorious tuna-dolphin trade case.[19] But labels are not always a clear winner. They require credible

---

[19] The United States banned the import of tuna from Mexico and other countries on the grounds that foreign fishing methods lead to "excessive" dolphin kills. Mexico took

certification and that costs money. The more complicated the supply chain, the more money it costs to trace through the environmental consequences of producing the product, including direct and indirect inputs. For carbon labeling, this requires information on energy use and emissions intensity at all points in the supply chain, from inputs to processing to storage to shipping to final sale. For a full carbon footprint, energy used in disposal and recycling opportunities would be relevant.[20] And the more complicated the supply chain, the easier it is to bias the carbon accounting to favor domestic production over imports. The relative simplicity of supply chains in agriculture helps explain why they are often used for illustration.

Trade economists have expressed concern that carbon labeling will put developing country exporters at a competitive disadvantage. There are three reasons. First, the additional cost of accounting and certification would weigh more heavily on small, distant suppliers to the market. Brenton et al. (2009) suggest that the cost of meeting strict standards imposed by agro-food industries in Europe may be a major reason for the marginalization of small farmers from horticultural export markets in the past two decades. Second, the accounting may intentionally or unintentionally be biased against these exporters and in favor of domestic producers. The shortcut of using emission parameters derived from industrial countries without correct adjustment for differences in production technology is an example. A third concern is that voluntary carbon labeling can slip into mandatory requirements for government procurement, or mandates for regulated industries. Renewable energy mandates and the strength of the corn ethanol industry in the United States, together with harsh treatment of imported ethanol made from sugarcane, reveal the protectionist potential.

The now-discredited concept of food miles illustrates some of the complexities. The simple and superficially appealing notion is that food grown or raised close to the table where it is consumed would reduce emissions from transportation. So it would, if all else were equal. All else is not equal, however, and a mere counting of food miles conveys

the case to the GATT and prevailed. Some have argued that a U.S. labeling requirement would have been legal and less intrusive.

[20] The common term is carbon labeling, although in principle – and sometimes in practice – all greenhouse gases are included.

no information on the type of transportation used. As noted earlier, emissions per ton mile can vary by a factor of 100 as between sea transport and air freight. Moreover, transportation is only one component of the carbon emissions profile. Mechanization of planting and harvest, irrigation, and the use of fertilizers also contribute. Brenton cites a study showing that cut roses produced in Kenya and air-freighted to Europe produced fewer emissions than roses grown in the Netherlands. Solar energy in Kenya undercuts fossil fuel energy in the Netherlands, despite the air freight handicap. Next, although important, carbon is only one of several greenhouse gases that affect climate. This may be especially important in agriculture where nitrous oxide from soils and methane from ruminants are involved. Carbon emissions are an incomplete global warming criterion. Finally, the ability of consumers to absorb and act on multi-criteria labels can be pushed too far. Consumers may have good intentions with regard to climate change, but they may also have good intentions concerning the use and abuse of pesticides, payment of fair wages, the use of child labor, gender discrimination, recyclability, and many other product features. Wrapping these preferences up into a simple and accurate labeling system remains a challenge.

In addition to the labeling issue, global warming policies offer fertile ground for other trade disputes. These include arcane rules concerning subsidies, technical barriers to trade, discriminatory government procurement, protection of intellectual property, non-discrimination and national treatment, and others. Some, such as clashing subsidies for green technology, are quite interesting and may have political consequences. But all of these disputes have been with us for many decades, and global warming policy poses no novel challenges.

Carbon Embodied in Trade

Trade allows a country to consume a bundle of goods and services different from the bundle it produces. The amount of carbon and other greenhouse gases embodied in the consumption and production bundles will differ. This has attracted efforts to determine the carbon balance of trade.

The UNFCCC carbon accounting methodology is based on a production or territorial concept – it attributes all emissions released within the territory of a country to that country.[21] An alternative would be to estimate total emissions associated with the consumption of goods and services by residents of a country. The principle adjustments in moving from a production- to a consumption-based accounting is to add emissions associated with the production of imported goods and subtract the emissions associated with the country's exports. The difference between emissions estimated on a production versus a consumption basis is sometimes called the carbon balance of trade. It will be positive if the emissions embedded in exports exceed emissions embedded in imports. The immediate determinants of a country's carbon balance are the level of exports relative to imports and the composition and carbon intensity of the country's trade.[22] There is no presumption that a positive carbon balance of trade is either desirable or undesirable.

A full calculation of carbon balances is a data-intensive exercise involving national input-output tables, estimates of CO_2 emissions factors by industry, and trade data by industry and by source and destination. A significant challenge is dealing with imports for processing and re-export. The results can be quite interesting. Helm et al. (2007), using admittedly rough estimates, calculate that by 2006, the UK trade deficit in greenhouse-gas emissions was 341 million metric tons CO_2e, or about 50 percent of total UK domestic greenhouse-gas emissions. He also estimated the net carbon deficit for the United Kingdom arising from one sector – international tourism – to have increased more than five-fold between 1990 and 2003.[23] "Sunshine tourism" may be at work here.

In the past thirty years, China has emerged as "factory to the world." Pan and his colleagues (2008) set out to measure China's carbon trade balance where the balance is defined as the difference in emissions under the production and consumption accounting schemes. They find that in 2006, emissions associated with Chinese production were 5,500 million tons CO_2, and that emissions associated with

---

[21] UNFCCC does not allocate the substantial emissions from aviation and shipping by country, but reports them as a memorandum item.

[22] Helm (2007) describes a third accounting scheme based on the nationality of emitters, not their spatial location. It has a parallel with the distinction between Gross National Product and Gross Domestic Product, but it is not clear why it would be useful for policy. Helm (2009) also considers this issue.

[23] Estimates based on greenhouse gas intensities by country, not on actual data on emissions per tourist or per tourist dollar spent.

Chinese consumption were 3,840 million tons, giving a carbon emissions surplus of 1,660 million tons. Switching from a production-based to a consumption-based accounting scheme would reduce China's emission growth rate from 12.5 to 8.7 percent annually for the period between 2001 and 2006. Nakano et al. (2009) undertook an ambitious study covering forty-one countries and regions (including 90 percent of world GDP) and seventeen industries. Their principal findings were: (1) between 1995 and 2000, more than half of the global increase in consumption-based CO_2 emissions took place within the OECD, but only one-third the increase in production-based emissions took place there[24]; (2) six OECD countries (France, Germany, Italy, Japan, United Kingdom, United States) accounted for 91 percent of the world's carbon trade deficit in 2000; (3) five non-OECD countries (Russia, China, Indonesia, India, and South Africa) account for 75 percent of the world's carbon trade surplus.

There is no need to pile up further estimates. The point is clear. International trade drives a wedge between where a product is produced and where it is consumed. Trade in goods reflects embodied emissions just as it reflects embodied labor. The relevant questions are: So what? Why should we care? Obviously, if countries are going to use border adjustments to protect competitive position, the embodied carbon content of particular products would have to be calculated. But this does not explain why the carbon balance of trade need be calculated. Studies of the *labor* content of trade have a very dubious history. Are carbon balance studies equally dubious? Certainly *bilateral* carbon balance studies are questionable. Trade economists have gone to great length to point out that bilateral trade balances have little economic merit and bilateral labor balances even less. If a country's carbon balance has any meaning, it is its global balance. But even that is suspect. We do not look to a banana trade balance. Why should we look for a carbon balance?

The answer lies mainly in the incomplete coverage of the Kyoto Protocol. With only a limited number of countries undertaking emission reduction obligations, both the global and national reduction targets are compromised by changing trade patterns in carbon-intensive

---

[24] This should not be taken as evidence of carbon leakage as discussed earlier, as restrictions on carbon emissions in this period were absent or very minimal.

products. A shift in production and trade of these products from restricted to non-restricted countries undermines the integrity of the original targets. More specifically, under the Kyoto accounting system, trade shifts tend to overstate the emission reduction accomplishments of Annex 1 countries and exaggerate the emissions levels of non-Annex 1 countries. Measuring changes in carbon trade balances provides a more accurate picture. Moreover, moving beyond Kyoto, the *construction* of a global cap-and-trade system would require consumption-based carbon emissions data that accounted for trade in order to obtain equity in the international distribution of emissions permits. Data on emissions measured on both a production and consumption basis are needed – precisely what the carbon balance provides. In principle, if the carbon content of current trade is considered in determining an equitable distribution of permits, historic trade patterns might also be examined – another invitation for disputes. Once a universal cap-and-trade system were in place, however, carbon trade balances would automatically seek efficient levels, and their calculation become unnecessary.[25] Until that day arrives, however, it is useful to have both production- and consumption-based accounts to accurately measure emission reduction efforts.[26]

The three previously cited studies illustrate the positive role that carbon trade accounting can play. The Helm study illustrates that what appeared to be an exemplary effort in the United Kingdom to reduce greenhouse-gas emissions was in fact the result of the United Kingdom's "dash for gas" as an energy source, and its systematic de-industrialization coupled with rapid increase of imports of carbon-intensive products, neither of which was primarily motivated by global warming concerns. In this instance, consumption-based accounting would have been more accurate and insightful than production-based accounting. The China study is important as it draws attention to the need to acknowledge trade patterns if and when a global carbon budget

---

[25] Presumably, countries with positive carbon balances such as China would be "compensated" for their trade role by receiving a disproportionate share of emissions permits.

[26] One could also fault the specific national-level Kyoto targets, which are expressed as reductions in national production of emissions, as having little relation to national consumption emissions. This would be correct, but the actual allocation of emissions reduction targets in Kyoto had no discernable pattern and appears arbitrary.

is established and countries such as China negotiate a fair and efficient share in a global cap. The Nankano study confirms the dynamic nature of international trade, as well as the need to accommodate major shifts in trade when setting emission targets and when allocating emissions permits.

Global Warming Policy and the Dutch Disease

The story line for climate policy revolves around an increase in the price of carbon. It follows that the cost and price of carbon-intensive goods will rise. In a globally efficient situation, the price of carbon will be equal across borders. The price of carbon-intensive goods will increase relative to other goods and services, and the composition of world production will tend to shift away from carbon-intensive goods. At the same time, the use of low-carbon or carbon-free energy sources will be encouraged. Provided that the price of carbon reflects the present value of the sum of the stream of marginal damages from emissions, world welfare is increased. The burden of higher carbon prices will be felt in the first instance by fossil fuel suppliers, who will see the terms of trade turn against them, and by net exporters of carbon-intensive goods, who will face shrinking markets. The potential loss of production and exports by countries that rely on carbon-intensive activities may limit their willingness to restrict emissions. As explained later, if emission-intensive exporters also become net sellers of emissions permits in a cap-and-trade system or receive substantial transfers in a harmonized tax system, they may also catch the "Dutch disease" adding to their reluctance to restrict emissions.

The "Dutch disease" is an affliction arising from good fortune. If a country enjoys a rapid, sustained increase in the export of a particular commodity (often a natural resource), the increased revenue increases aggregate demand for both traded and non-traded goods. Increased demand for tradeables is met by imports, but demand for non-traded goods requires a shift in the structure of production away from traded goods. This can be accomplished by allowing the exchange rate to appreciate, which increases the relative price on non-tradeables and draws resources (e.g., labor) away from the (tradeable) manufactures sector. That sector contracts. In itself, this adjustment is not necessarily damaging unless one makes further assumptions that the manufactures export sector is particularly well suited to provide jobs, or that there are externalities associated with manufactures exports. These assumptions concerning the central role of industrialization and manufacturing in development are quite common.[27]

Revenues from the sale of emission rights can be fitted into this framework. The revenues may boost demand, appreciate the currency, and lead to a contraction of production and export of manufactures. Countries that produce and export energy-intensive manufactures and that may receive relatively large revenues from the sale of emission rights appear especially vulnerable – first to the decreased demand for energy-intensive manufactures, and second, working through currency appreciation and the Dutch disease adjustment process, to the contraction of the manufactures sector and an additional decline in all manufactures exports.

As a first step in assessing the Dutch disease, it is useful to consider the amounts of money that permit trade could conceivably generate. Although unlikely in the near term, the amounts that could cross borders are potentially very large. The International Monetary Fund (IMF) reports that using one general equilibrium model (G-cubed), China could receive between 5.5 and 5.9 percent of its GDP in 2040 from the international sale of emission rights, depending on whether emission rights are allocated by population or proportional to initial levels.[28] Using a different model, assuming an aggressive global target of 450 ppm for CO_2, and assuming emission rights are allocated by population, India may enjoy net receipts from international permit sales exceeding 4 percent of its GDP by 2020. These numbers are perhaps implausibly high, but illustrate the potential fiscal impact of a global cap-and-trade system.

---

[27] The Dutch disease might be compared to the classic transfer problem, which involved adjustments in the country in which the transfers originated (e.g., German repatriations following World War I). See the debate between Keynes and Ohlin in volume 39 of *The Economic Journal* (1929). The Dutch disease concerns adjustments in the recipient country.

[28] IMF (2008). The IMF is careful to point out that results are model-specific. China and India are expected to become net sellers of permits because they are known to have relatively low marginal abatement costs.

The implications have been investigated by Mattoo and his colleagues (2009a). They use a multi-sector, multi-country general equilibrium model to trace out welfare, output, and export effects on different countries and country groups in 2020 of various emission reduction targets. The central reduction targets are a 30 percent cut below 2005 levels for rich countries and a 30 percent cut below business-as-usual (BAU) levels for poor countries (which would allow growth of emissions if BAU levels are set sufficiently high). The analysis is restricted to carbon released from fossil fuels. The two scenarios of interest to us are when there is no permit trade among regions and when a full global cap-and-trade system is in place. The latter, of course, equalizes carbon prices internationally and sets up the financial transfers that may lead to the Dutch disease effects. Three countries are of particular interest: China and India, representing carbon-intensive economies; and Brazil whose carbon intensity is only one-third the average of low- and middle-income countries.[29]

The results show that without permit trading, manufacturing contracts 2.9 percent in China, 3.7 percent in India, and 0.8 percent in Brazil. With emissions trading, which sets financial flows and currency changes in motion, manufacturing output declines by 6.5 percent in China and 5.5 percent in India, but gains 0.6 percent in Brazil. Manufactures exports tell much the same story. In China, they shrink by 4.5 percent without trading and by 9.4 percent with trading; in India, they shrink 7.3 percent without trading and 10.7 percent with trading: in contrast, in Brazil they shrink 1.5 percent without trading but gain 2.7 percent with trading. The message is that countries contemplating manufactures-led export growth may have legitimate concerns about partnering in a global cap-and-trade system.[30]

These results are not cast in stone. Some countries may welcome a shift away from traditional carbon-intensive manufacturing and use emissions reduction receipts to fund ambitious and expensive transformation into a higher-technology production structure. Some

---

[29] Carbon intensity is measured as tons per million dollars of output. The figures are 772 (China), 535 (India), and 149 (Brazil). High-income countries average 109.

[30] Burniaux et al. (2009) found evidence of the Dutch disease reducing welfare in some Eastern European countries if there were full emissions trading among Annex 1 countries. They attribute the welfare loss to premature scrapping of capital in manufacturing.

may use the receipts to subsidize exports. And some may resist the appreciation of their currencies.

Manipulating Permit Markets

A full-blown global emissions trading system, complete with a cap and an agreement on permit allocations, makes a convenient analytical construct but is unlikely in the near term. Gradual linking of national and regional schemes is not inevitable but is more likely. The greater the disparity in carbon prices before linking, the greater the potential efficiency gains. In principle, greater efficiency should support a stronger emission reduction effort. But this may not turn out to be the case.[31]

International linking of national cap-and-trade systems is vulnerable to various manipulations. Following well-known mercantilist tendencies, governments may prefer to become permit exporters rather than importers in a linked cap-and-trade system. Firms may also lobby to keep permit prices low. Because enlargement of the cap *appears* to be a costless step (a stroke of the officials' pen), governments may therefore take the opportunity of linking to increase their caps. Governments in prospective permit-selling countries may seek to increase their receipts; governments in the buying countries may seek to reduce their purchases. In short, the linking may change the political economy dynamics. Carbon prices would still be equalized, but emissions would rise rather than fall.

Other opportunities for manipulation are familiar from trade theory. Linking creates value for the buyer and seller, but the distribution of the gains from permit trade – the terms of trade – is up for grabs. The full array of trade policy instruments can be brought into play to tilt the terms of trade: tariffs, quotas, tariff-rate quotas, discounts on imported emissions permits, or "minimum domestic requirements" – all can be deployed by permit-importing countries. A similar array of tools is available to exporting countries. It would be tedious to analyze all of them. In general, however, the effect of these restrictions on trade in permits is to reduce welfare, and their use is likely to increase global

---

[31] See Burniaux et al. (2009), Helm (2003), Jaffe and Stavins (2010), and Rehdanz and Tol (2005).

warming. The WTO, which regulates trade in goods and services, may need to consider regulating trade in emissions permits.

Conclusions

The most general conclusion is that the introduction or liberalization of trade will improve welfare *if* optimal climate policies are followed. At the same time, the introduction of climate policies may change the level, gains, and composition of trade, but free trade remains the optimal trade policy. A second conclusion is that concern for competitive losses is almost universal – rich countries fear carbon leakage, fossil-fuel-exporting countries fear terms-of-trade effects, and developing countries fear some combination of border adjustments, carbon-related non-tariff trade barriers, and a possible bout of the Dutch disease. These fears may be exaggerated but there is enough validity to threaten climate negotiations. A third conclusion is that many of the trade concerns arise from the partial and incomplete nature of restrictions on greenhouse-gas emissions. These concerns would ease if a global cap-and-trade or tax system were put in place.

References

Aldy, J. and W. Pizer (2009). The Competitiveness Impacts of Climate Change Mitigation Policy. *Pew Center on Global Climate Change.* Arlington, VA.

Babiker, M. (2005). Climate Change Policy, Market Structure, and Carbon Leakage. *Journal of International Economics* 65: 421–45.

Barker, T., S. Junankar, H. Pollitt, and P. Summerton (2007). Carbon Leakage from Unilateral Environmental Tax Reforms in Europe, 1995–2005. *Energy Policy* 35: 6281–92.

Brenton, P., G. Edward-Jones, and M. Jensen (2009). Carbon Labeling and Low-Income Country Exports: A Review of Development Issues. *Development Policy Review* 27 (3): 243–67.

Burniaux, J., J. Chateau, R. Dellink, R. Duvall, and S. Jamet (2009). The Economics of Climate Change Mitigation: How to Build the Necessary Global Action in a Cost-Effective Manner. *OECD Economics Department Working Paper 701.*

Chichilnisky, G. (1994). North South Trade and the Global Environment. *The American Economic Review* 84 (4): 443–9.

Copeland, B. and M. S. Taylor (1995). Trade and Transboundary Pollution. *The American Economic Review* 85 (4): 716–37.

(2003). *Trade and the Environment: Theory and Evidence.* Princeton, NJ: Princeton University Press.

(2005). A Trade Theory View of the Kyoto Protocol. *Journal of Environmental Economics and Management* 49: 205–34.

Dean, J. (2002). Does Trade Liberalization Harm the Environment? A New Test. *The Canadian Journal of Economics* 35 (4): 819–42.

Dean, J., M. Lovely, and H. Wang (2009). Are Foreign Investors Attracted to Weak Environmental Regulations? Evaluating the Evidence from China. *Journal of Development Economics* 90: 1–13.

Glombek, R. and M. Hoel (2004). Unilateral Reductions and Cross-Country Technological Spillovers: Advances in Economic Policy and Analysis. *The BE Journals of Economic Analysis and Policy* 4 (2): Article 3.

Goulder, L., M. Hafstead, and M. Dworsky (2009). Impacts of Alternative Emissions Allowances Allocation Methods under a Federal Cap- and-Trade Program. *NBER Working Paper 15293.*

Grossman, G. and A. Krueger (1993). Environmental Impacts of a North American Free Trade Agreement. In *The Mexico-U.S. Free Trade Agreement*, P.Garber (ed.). Cambridge, MA: MIT Press.

Helm, C. (2003). International Allowances Trading with Endogenous Allowance Choices. *Journal of Public Economics* 87: 2737–47.

Helm, D. (2009). Climate Change Policy: Why Has So Little Been Achieved? In *The Economics and Politics of Climate Change*, D. Helm and C. Hepburn (eds.). Oxford: Oxford University Press.

Helm, D., R. Smale, and J. Phillips (2007). *Too Good to Be True? The UK's Climate Change Record.* Available at http://www.dieterhelm.co.uk/

Ho, M., R. Morgenstern, and J.-S. Shih (2008). Impact of Carbon Price Policies on U.S. Industry. *Resources for the Future Discussion Paper* 08–37.

Hufbauer, G., S. Charnovitz, and J. Kim (2009). *Global Warming and the World Trade System.* Washington, DC: The Peterson Institute for International Economics.

Hummels, D. (2009). How Further Trade Liberalization Would Change Greenhouse-Gas Emissions from International Freight Transport. Presented at OECD Global Forum on Trade and Climate Change, Paris, June 9–10, 2009.

International Monetary Fund (2008). *The Fiscal Implications of Climate Change.* Washington, DC: Fiscal Affairs Department, IMF.

Jaffe, J. and R. Stavins (2010). Linking of Tradeable Permit Systems in International Climate Policy. In *Post-Kyoto International Climate Policy*, J. Aldy and R. Stavins (eds.). New York: Cambridge University Press.

Mattoo, A., A. Subramanian, D. van der Mensbrugghe, and J. He (2009a). Can Global De-Carbonization Inhibit Developing Country Industrialization? *World Bank Policy Working Paper 5121.*

(2009b). Reconciling Climate Change and Trade Policy. *World Bank Policy Working Paper 5123*.

Nakano, S., A. Okamura, N. Sakurai, M. Suzuki, Y. Tojo, and M. Yomano (2009). The Measurement of CO_2 Embodiments in International Trade: Evidence from Harmonized Input-Output and Bilateral Trade Databases. *OECD STI Working Paper 2009/3*.

Pan, J., J. Phillips, and Y. Chen (2008). China's Balance of Emissions Embodied in Trade: Approaches to Measurement and Allocating International Responsibility. *Oxford Review of Economic Policy* 24 (2): 354–76.

Pearson, C. (2000). *Economics and the Global Environment*. Cambridge: Cambridge University Press.

(2004). *United States Trade Policy: A Work In Progress*. Hoboken, NJ: Wiley.

Rehdanz, K. and R. S. J. Tol (2005). Unilateral Regulation of Bilateral Trade in Greenhouse Gas Emissions Permits. *Ecological Economics* 54: 397–416.

Reinaud, J. (2008). Issues Behind Competitiveness and Carbon Leakage: Focus on Heavy Industry. *OECD/IEA Information Paper, October 2008*.

United States Government (2009). The Effects of HR 2454 on International Competitiveness and Emissions Leakage in Energy-Intensive Trade-Exposed Industries. *Interagency Report, December 2*.

WTO-UNEP (2009). *Trade and Climate Change*. Geneva: WTO.

8

The Challenge of International Cooperation

Introduction

The laws of physics give de facto property rights to greenhouse gas emitters. The doctrine of state sovereignty leaves control up to individual countries. Although the elegant Coase Theorem suggests that this does not have to be a barrier to negotiating efficient levels of mitigation, the international market for cooperation appears dysfunctional.[1] This chapter provides some partial answers as to why this is so. The subsequent chapter assesses current climate negotiations.

Climate change is a global "public bad" and preventing climate change is a global "public good."[2] We start with a brief conceptual discussion of these terms. This leads to the conclusion that coordinated international action, presumably through a post-Kyoto international environmental agreement (IEA), is needed. But it also highlights the serious difficulties involved. The next section considers the objectives, timing, and especially the participation in such an agreement. This may be thought of as the willingness to cooperate. A combination of carrots and sticks may encourage that willingness. The necessity for cooperation arises from the need to minimize the cost of greenhouse

---

[1] The Coase Theorem demonstrates that with clear property rights and minimal transactions costs, the parties to an externality can reach an efficient solution regardless of who is awarded the rights.

[2] If, as expected, the costs of adjusting to a new climate exceed the benefits, change is a public bad. Costs are best viewed as adjustment costs. There is no clear evidence that after sufficient millennia have passed and all environmental, economic, and social adjustments to higher temperatures have been made, human well-being will be higher or lower than today.

gas abatement and the rapidly closing window for aggressive climate policy. The case for inclusion of medium and large developing countries (and the United States) in mitigation efforts is compelling. Even with carrots and sticks, however, there appears to be a mismatch between the need for and the willingness to cooperate.[3]

Concepts: Public Goods, Public Bads

The two essential characteristics of pure public goods are well known. First, they exhibit non-excludability. Once produced, it is technically difficult or costly to exclude anyone from using them, or the services they provide. Second, they exhibit non-rivalry. One person's consumption of the good does not diminish the amount available for others. Knowledge – say, the invention of the calculus – is a good example.[4] Once known, it is difficult to prevent its use. And my use of the calculus does not preempt your use. A public bad, which diminishes utility or well-being, has the same two characteristics.

Global public goods are simply public goods enjoyed by many countries. Climate is an example. Whether one chooses to consider the public good (preventing climate change) or the public bad (climate change) is immaterial. Once climate change occurs, it is non-excludable. The citizens of India or any other country cannot choose to live in the climate of the past (non-excludability). And the costs of adjusting to climate change that are borne by one country in no way limit adjustment costs borne by other countries (non-rivalry).

The economic problem posed by public goods is straightforward. Left to themselves, private markets will undersupply them. Firms will not find it profitable to produce a public good or service because, with non-excludability, no one will be willing to buy what they can have for free. Additionally, non-rivalry in consumption supports a zero price. Once produced, it would be inefficient to limit access by charging a price. It follows that social welfare can be increased if government undertakes the production of public goods or subsidizes the production in the private sector. By the same logic, an unregulated private sector

---

[3] This chapter is concerned with mitigation efforts – a public good. Adaptation measures, even if partially funded internationally, are not a global public good. Adaptation is discussed in Chapter 5.

[4] The transmission of knowledge – education – is both excludable and rivalrous.

will overproduce a public bad because it has a zero cost to those who produce it. Global warming fits this situation neatly. Until very recently, greenhouse gas emissions – an externality from production and consumption – have been unpriced and too much has been emitted. Just as governments can promote the production of public goods, they can curtail public bads and improve social welfare. The obvious action is to attach an implicit or explicit cost (price) to the activity that produces it. As explained later in the chapter, providing *global* public goods present a more difficult problem, as there is no international government with coercive or taxing powers, and provision of the global good (restriction of the global bad) is the result of international negotiations.[5]

Determining the optimal supply of a public good is complicated, for two reasons. First, because of non-excludability, there is no conventional market and market price to reflect demand and value. Individuals may be asked what value they place on the good but they have little incentive to provide a true answer. The strategic response is to lowball the estimate if they think they will have to pay and exaggerate it if others are expected to pay. In the context of global warming, in which the agents are governments, not individuals, a similar problem can arise. Countries may be unwilling to reveal their true "demand" for slowing climate change, either exaggerating their benefits if they expect others, not themselves, will pay or minimizing their benefits if they expect to pay what they reveal, while counting on others to pick up the tab. Genuine uncertainty concerning national-level costs and especially benefits, as well as genuine differences concerning discount rates complicate the separation of strategic from valid claims. Still, benefit functions (damages averted) are in principle estimates of objective phenomena – coastal flooding, crop yields, disease, and so on – so that the scope for misstatement may be limited. Such strategic behavior is considered later in the chapter.[6]

---

[5] Nordhaus (2007) has noted that the 1648 Treaty of Westphalia established that obligations can only be imposed on a sovereign state with its consent. This, however, is only part of the legal story. The 1944 Trail Smelter case and subsequent United Nations declarations restrict the right of a state to create environmental damages in other states or areas beyond its national jurisdiction.

[6] Uncertainty concerning true costs and benefits is sometimes used to support proportional reductions by all polluters. This would be highly inefficient and costly in the case of global warming because marginal abatement costs are known to differ substantially.

The second complication is that public goods are supplied in fixed quantities and, unlike private goods, individuals (countries) cannot adjust the quantity they receive (Heal 1998). This is an important point as it links efficiency and questions of equity. An equilibrium reached in a competitive private goods economy is efficient because individuals can adjust the quantities of goods they buy to equate the ratio of marginal utilities they receive to the relative prices of the goods. Differences in income and preferences among individuals are accommodated by differences in quantities purchased. Efficiency is achieved through the market and is independent of the initial distribution of the goods. This does not work with public goods. They cannot be traded to accommodate differences in income and preferences – for better or worse, one size will have to fit all. How can the supply of a public good – for example, preventing global warming – be made to fit the needs of China and Australia, Chad and Saudi Arabia, South Korea and Russia?

One solution involves decoupling the distribution of abatement *effort* from the distribution of abatement *financing*. The first step uses a uniform carbon tax or a cap-and-trade system to ensure that marginal abatement costs are equal across countries. The tax or cap can be set either to meet a temperature or concentration target, or set so that global marginal costs are equal to global marginal benefits (both discounted). The equality of marginal costs ensures an efficient allocation of abatement effort. Financing the aggregate cost of abatement is then allocated among countries so that the share each country pays reflects their marginal willingness to pay for the global supply of the public good – in this case, the reduction in greenhouse gas emissions. In other words, instead of adjusting quantity purchased to price, the price a country pays is adjusted to the fixed quantity it receives. These are known as Lindahl prices. The result is Pareto optimal – the public good has been fully funded, and each individual country is satisfied with the amount provided. Efficiency is achieved but only with the appropriate distribution of abatement costs.

Willingness to pay follows the benefit principle of public finance. Allocating financing shares on the basis of willingness to pay means that countries facing high global-warming damage costs will face higher abatement costs. But equity is also served because willingness to pay is itself determined in part by income and preferences. *Ceteris paribus*, poor countries will pay a lower share than rich countries. In other words, the ability-to-pay criterion of public finance is accommodated (Buchholtz and Peters, 2007). The final step is a system of transfers from countries whose share of total abatement costs as describe earlier exceeds their share of actual abatement expenditures, to countries whose actual expenditures exceed their share. In a cap-and-trade system, this transfer is implicit in the allocation of emission permits among countries. Fiscal transfers would need to be made in a carbon tax system.

Lindahl prices are not a panacea, however. They do not resolve legitimate uncertainty concerning abatement cost and benefit functions, and do not prevent strategic misrepresentations. Nor do they eliminate the need for transfers if national-level cost and benefit functions differ among countries. Negotiations may get hung up on other criteria, for example demands for equal emissions per capita, in which case, least cost abatement cannot be attained. "Historic responsibility" may be invoked to shift the abatement cost burden toward the North. However, some northward shift of the abatement cost burden is already implicit in the willingness-to-pay criterion. A lower marginal utility of income in rich countries suggests that permits in a cap-and-trade system should be disproportionately given to poor countries (Chichilnisky and Heal 1994; Sheeran 2006; see also Chapter 3).

Transfers are an important consideration. As previously suggested, one justification for a biased distribution of permits, or tax transfers, is efficiency in the face of different marginal utilities of income. It is not inconsistent with biased distribution that is based on notions of equity and on the North's historic responsibility for creating the global warming problem.[7] So far so good. The peculiar nature of public goods, and the great disparities in the world's per-capita income, support shifting the abatement cost burden toward rich countries on efficiency grounds; historic responsibility supports the same shift; the ethical ability-to-pay criterion supports the shift; the Polluter Pays Principle supports the shift, at least until per-capita emissions are

---

[7] If some cost allocations are more efficient than others, negotiations are no longer a zero-sum game. In principle, negotiating distribution in a positive-sum game should be easier.

equalized between rich and poor; the UN Framework Convention on Climate Change promotes it.[8]

But there is a major catch. Lindahl prices apply to situations where a government is there to provide the optimal quantity of the public good and then has to decide how to allocate the costs. These situations do not confront the constraints of sovereignty and self-interest in negotiations. Lindahl prices do not speak to a central problem in global warming negotiations – free-riding. They do not address whether cost distribution schemes can be used to increase cooperation and participation. The following section reports a number of studies where the efficient allocation as described earlier conflicts with incentives for broad participation – another efficiency objective. As it turns out, there is a conflict between the efficient distribution of abatement costs and distributions that encourage widespread participation. We return to this issue after discussing participation incentives.

More broadly, the connection between efficiency and distribution (equity) in the analysis of a public good such as global warming policy is critical. Not only are distribution issues central in an international political system that relies on negotiation rather than coercion, but for public goods supplied in fixed quantities, the optimal supply itself depends on who bears the cost.

Supplying Global Public Goods

An Ecological Surplus

Global warming is accurately portrayed as a public bad and its prevention as a public good.[9] Warming results from a negative externality – the costs of greenhouse gas emissions are not fully borne by emitting countries. As with any uncorrected externality, an improvement in

---

[8] If transfers through a cap-and-trade or carbon tax system are not available, we are in a second-best world where efficiency from equalized marginal costs may conflict with the efficiency obtained by the broadest possible participation in a global IEA. Shiell (2003) argues that the initial allocations of permits in a cap-and-trade affect transfers, but he does not show that efficient transfers would provide incentives to join anIEA. In fact, biased allocation of permits to poor countries could create implausibly large permit trading and rejection by rich countries. See page__ in this chapter.

[9] For thoughtful analysis of types and examples of global public goods, see Barrett (2007).

welfare may be within reach. International environmental agreements (IEAs) to prevent global warming are motivated by this potential for welfare improvement. However, unlike domestic public goods, global public goods are the result of negotiations, not government fiat. This imposes a major constraint – absent coercion, all parties to an agreement must decide they are better off participating rather than abstaining.

Cooperation among countries in supplying global public goods rests on welfare gains derived from greater efficiency. In the absence of cooperation, countries will presumably pursue their narrow self-interest.[10] For global warming, this implies pursuing abatement to the point where a country's marginal costs are equal to the marginal benefits it receives from its own actions (the non-cooperative solution). That level will be suboptimal as no account is taken of the benefits conferred on other countries. This, too, is best thought of in externality terms. Abatement by one country creates an external *benefit* for all other countries by reducing emissions and global warming. Unless countries are compensated for the external benefits they confer on others, too little abatement will be undertaken. In the absence of a supra-national environmental protection agency with the authority to compel action, the public good – abatement – will be undersupplied. The problem of undersupply provides the rationale for the Global Environmental Facility that helps fund projects in developing countries that create international externality benefits.

The overarching question is whether a climate IEA can be negotiated that would capture the efficiency and welfare gains from internalizing this externality, while respecting *both* a no-coercion negotiating constraint and differences in marginal utility of consumption between rich and poor. The problem can be framed as either internalizing a negative externality associated with greenhouse gas emissions, or a positive

---

[10] Maybe. The theoretical literature on IEAs pays almost no attention to domestic political considerations (except for trade-competitive concerns), and their impact on shaping policy. Additionally, some countries may wish to set a good example by abating beyond their narrow self-interest, hoping that others will follow. This can backfire, however. The additional abatement by the lead country decreases the incentive for others to abate by decreasing marginal benefits that others receive from their own abatement effort. Indeed, public-spirited action by one country can result in higher global emission levels (Hoel 1991). This perverse outcome is independent of possible carbon leakage.

externality associated with abatement, in a non-coercive, international negotiating context comprised of rich and poor countries. Notice that it is the inability to exclude non-participating countries from enjoying the benefits of an IEA that distinguishes global warming from international trade liberalization. WTO members can and do withhold Most Favored Nation tariff treatment from non-members.

For the moment, we set aside differences among countries in marginal utility of income and consumption so we can concentrate on participation in an IEA. The efficient solution that would maximize global welfare requires each country to abate to the point where its marginal abatement cost is equal to the *global* marginal benefit of its abatement.[11] Marginal (but not total) abatement costs among countries are then equal. Uniform carbon taxes or a cap-and-trade system can in principle accomplish this.[12] Efficiency does *not* require that the pattern of payment for abatement – the financing – replicate the pattern of abatement effort. In general, transfers or side payments will be consistent with efficiency in the allocation of abatement effort. Transfers can also be used to increase participation in an IEA. The welfare gain from moving from the non-cooperative to a cooperative solution, internalizing at least some of the externality associated with abatement, is the rationale for an IEA. Chander and Tulkens (1992) call this the ecological surplus, comparable to the gains from trade in voluntary exchanges.

The emergence of an IEA that captures a large share of the surplus is not guaranteed even if the potential gain from cooperation is large. An IEA that merely requires members to do what they would normally do is ineffective.[13] An IEA that falls short of equalizing marginal abatement costs from its member countries (as the Kyoto Protocol did) does not obtain the maximum possible surplus. Most importantly, an IEA that only enrolls a small number of countries only captures

---

[11] For the visually oriented, recall that with public goods, marginal benefit curves are added vertically because of non-exhaustibility, and marginal cost curves are added horizontally.

[12] An *internationally* levied carbon tax need not be uniform if countries set domestic carbon taxes to equate their marginal costs with their own marginal benefits.

[13] A good example is the 1972 Ocean Dumping Convention (Pearson 1975). Except for a short list of prohibited materials, countries were pretty much free to determine what and where they dump.

a small fraction of potential welfare gains. Participation by countries that have low abatement costs and large emissions is the key.

The main questions studied in the theoretical IEA literature as it relates to global warming are the following:

1. Under what conditions are IEAs stable? What determines the decision to join? What determines the decision to defect? What are the incentives to free ride?

2. Are the potential welfare gains from cooperation large or small? Do stable coalitions capture a large or small share of the potential gains? Are the emission reductions (the environmental benefit) attributable to the IEA significant?

3. What is the role for side payments (transfers)? Do side payments encourage participation and stability? What rules should govern the allocation of abatement costs and the distribution of the IEA surplus?

4. What is the role for sanctions, especially trade sanctions, to encourage participation and discourage defections?

Self-Enforcing IEAs

A good starting point for understanding the difficulties of negotiating an effective agreement is the concept of self-enforcing IEAs (Barrett 1994). If coercion is ruled out, participants in an IEA must find it in their interest to join and carry out its obligations.[14] For a stable agreement, non-members must find it in their interest to remain outside the IEA. This is sometimes termed internal and external stability. If unstable, an IEA will either expand or shrink until it reaches a stable size. Depending on modeling, a single defection could trigger complete collapse. One extreme is the inclusion of all countries – the so-called Grand Coalition – which maximizes the surplus from cooperation. At the other extreme, the IEA dissolves or fails to form and countries are assumed to maximize their parochial interests – the non-cooperative equilibrium.

There are two obstacles to full cooperation: free-riding and asymmetry among countries in abatement costs and benefits.[15] The creation

---

[14] Most international treaties allow for opting out. Thus defections are a concern.
[15] Free-riding behavior derives from the same motivation that drives cooperation by coalition members – the pursuit of self-interest. Still, the negative connotation

of a surplus is the incentive to join. But non-participants also share the surplus without incurring additional abatement costs. This is the problem of free-riders, and it can thwart a successful IEA. For analytical purposes, it is useful to start with the assumption that all countries are identical with respect to costs and benefits. Later we consider the implications of asymmetries in cost and benefit functions.

For an individual country, the value of free-riding increases as membership by others in an IEA increases.[16] The externality benefits conferred on a non-member increase with the size of the IEA. The incentive for a country to free-ride is also high when the ratio of the benefits it confers on others from abatement to its own benefit is large. This ratio is likely to be high for global warming because greenhouse gases are uniformly mixed in the atmosphere.[17] The ratio is also higher for small countries that are minor sources of emissions. At very high ratios, any abatement approaches pure international altruism. These incentives to free-ride tend to constrain an IEA below the grand coalition and below its potential for welfare improvement.

Barrett investigates self-enforcing (stable) IEAs in a model that assumes identical countries. His conclusions are sobering. The models show that high levels of participation in a self-enforcing IEA will obtain only when the difference in net benefits between the cooperative and non-cooperative outcomes is small (that is, for a weak agreement). The analysis also shows that when the difference is large, the IEA will consist of only a small number of countries. (For some parameter values, there is no self-enforcing IEA.) In both cases, the additional benefits created by the IEA over the non-cooperative outcome are small. The intuition behind the first conclusion is that it is relatively easy to sign agreements that do not call forth much more abatement effort than would be warranted on the basis of narrow self-interest. This would be the case if the ratio external to own benefits is low. Some observers have attributed the almost universal adherence – and hence "success" – of the Montreal Protocol limiting ozone-depleting substances to this feature.

---

attached to free-riding is deserved. The coalition members create external benefits; the free riding non-members create external costs.

[16] With asymmetric costs, the identity of the members also matters.

[17] This contrasts with sulfur dioxide emissions that generally have a higher fraction of local- or national-level damages.

The intuition behind the second conclusion is that when the difference between the cooperative and non-cooperative outcomes is large, the incentive to free-ride or to defect is also large, and, in this model, the inducement to join – or the sanction for defecting – is too weak to prevail. The incentive to free-ride is to gain the benefits of others' abatement without bearing the cost of joining the coalition. Deterring free-riding or defections is weak as the only lever is increases or decreases in abatement by the IEA members, as they maximize their collective net benefits in response to new members or the departure of defecting members.[18] The effects of these abatement adjustment actions by the coalition are diffused over all countries, and therefore their inducement or punishment potential toward a single country is small and weak.

Barrett's analysis also highlighted the potentially pernicious effect of leakage – as signatories to the IEA increase their abatement, non-signatories decrease their abatement effort. This follows directly from the model structure in which non-members choose their abatement levels to equate their marginal costs and benefits. As IEA abatement increases, the incentive to abate among non-members is weakened. Indeed, under certain parameter values, leakage fully offsets any incremental abatement by the coalition. In that event, the IEA is completely ineffective and presumable collapses. A self-enforcing IEA does not exist. Leakage is a central issue both for the environmental effectiveness of an IEA and for concerns about international trade competitive losses. It is considered in detail in Chapter 7.

On a more positive note, the model confirms that the smaller the number of countries involved, the more powerful the incentive to join and the more powerful the sanction for defecting. With smaller numbers, the ratio of external to own benefits falls as does the incentive to free-ride. The total number of countries discussing post-Kyoto arrangements is very large, but only six countries contribute 71 percent of global CO_2, and two countries (China, United States) account for 42 percent.[19] Whether global warming is best treated as a small or large number case is not clear.

---

[18] A defecting member would lose as the surplus created by the IEA would shrink, but would also be relieved of its IEA abatement obligations.

[19] Counting the EU as one unit. CO_2 from fossil fuels only. Data for 2006.

It is also unclear how much light the negative results of this particular model cast on the global-warming *problematique*. The model assumes countries have identical marginal cost and benefit functions, side payments (transfers) are ruled out, and only a weak sanction is employed. These are not good descriptions of current climate change negotiations. To move closer to global warming, we must consider heterogeneous countries; investigate transfers; and examine stronger sanctions on free-riding.

Carrots and the Search for Cooperation

At first glance, the introduction of side payments (transfers) would seem to improve the prospects for cooperation. Side payments could bribe additional members to join and might help retain wavering members. A moment's reflection, however, suggests this is not always the case. The model described earlier explains low levels of cooperation on the basis of free-riding and not on asymmetry in costs and benefits among countries. In this situation, a self-enforcing IEA must make the side payment out of its own surplus to attract a new member. This reduces the attractiveness of the IEA to its members, encouraging defections. With members paying the full cost of the bribe and only receiving a fraction of the addition to the surplus from the new member (non-members capturing the balance), there may be some churning of membership but little, if any, increase. A credible, irrevocable commitment to maintaining membership while expanding might help, but such a commitment is generally not possible.

The case for side payments to counter free-riding while assuming that all countries are identical is weak. Strong asymmetries among countries as to costs and benefits of abatement, which characterize global warming, may present a better justification.[20] Without transfers, developing countries would bear about two-thirds of the gross discounted costs of an efficient abatement program over the next century. This would be a hard sell indeed. Presumably in an IEA with transfers, payments would flow from countries that benefit greatly from abatement to countries with low marginal abatement costs, often (but not

---

[20] Barrett (2001) provides a theoretical model demonstrating that strong asymmetry with respect to net gains justifies a change in the rules of the game, and cooperation "for pay" can lead to greater participation.

always) developing countries.[21] Alternatively, if a global cap-and-trade system is under consideration, emission permits can be disproportionately allocated to countries facing low marginal abatement costs. Their large share of total abatement costs would be compensated by revenues from permit sales. Still, shifting the analytical focus to asymmetries in costs and benefits, and allowing for transfers through different sharing rules, does not guarantee the stability of coalitions nor overcome all free-rider difficulties.[22] In fact, the outlook for a reasonably comprehensive, stable IEA that delivers major welfare gains from cooperation is cloudy. There are several reasons. One is the persistence of free-rider behavior. A second is the size of the transfers that may be necessary. Worries about loss of trade competitiveness is a third. Three recent studies demonstrate these points.

Three IEA Simulations
The general approach is similar. A game theoretic analysis is combined with a dynamic Integrated Assessment Model to generate numerical simulations of economic, climate, and environmental damage estimates; identify stable coalitions; and evaluate transfer schemes.

Carraro et al. (2006) divide the world into six regions (United States, European Union, Japan, Former Soviet Union [FSU], China, and Rest of World [ROW], mainly developing countries), which allows for fifty-eight possible coalitions. The United States, Japan, and EU are assigned steep abatement cost functions due to their current high energy efficiency, the assumption being that further energy savings will be difficult and expensive. Benefit (i.e., damage avoidance) functions are steep for the EU (large economic assets at risk) and ROW (economic structure, especially agriculture, thought to be vulnerable to climate change). China is given relatively flat abatement

---

[21] Unidirectional externalities, which flow from upstream to downstream, are an extreme example of asymmetrical benefits. In the absence of coercion or international legal protection, a Coase type negotiation will improve global welfare, but the laws of gravity require the agreement to follow the victim pays principle. The pollution flows downstream and the transfer payments flow upstream.

[22] For analysis of the effects of transfers on participation in a global-warming IEA, see Barrett (2001), Eyckmans and Tulkens (2003), Carraro et al. (2006), and Eyckmans and Finus (2007). For an exposition of the game theoretic underpinnings of IEA formation see Finus (2001), and Bosetti et al. (2009) Annex 2 *Coalition Theory: A Selective Survey*.

cost and benefit functions. Welfare is measured as discounted global consumption integrated over the period between 1990 and 2300. The welfare increase (surplus) from cooperation in a particular coalition is measured as the percent of the surplus that would be obtained with full cooperation – the Grand Coalition – and thus ranges from zero (no cooperation) to 100 percent. Setting aside the Grand Coalition at 100 percent, the welfare gains for the top-ranked fourteen coalitions range from 99 percent down to 69 percent. China is identified as a member in all of the top twelve, confirming its importance. The United States is identified as a coalition member in eight of the top twelve.

While the numbers for the top coalitions would seem reassuring, there are two more disturbing results. First, the gain from full cooperation, while not trivial in absolute numbers at $771 billion, is only about 0.5 percent of total discounted global consumption over this period. From this perspective, cooperation, although desirable, does not appear essential.[23] The *environmental* benefits from cooperation measured by emission rates, atmospheric carbon concentrations, and temperature change, however, are far more impressive. Moving from no cooperation to full cooperation, cumulative carbon emissions over the same period are cut in half, and atmospheric concentrations in 2300 with full cooperation are only 42 percent of the no-cooperation level. The apparent discrepancy between the modest economic welfare and large environmental effects of cooperation arises because welfare benefits are discounted over a very long time period, and carbon emissions and concentrations are, quite appropriately, not discounted.

The more problematic conclusion is that in the absence of transfers, none of the fifty-eight potential coalitions, including the Grand Coalition, is stable. According to the rules of the game, without internal and external stability, there would be no coalition, no cooperation, and no IEA. Whereas many potential coalitions could increase welfare (create a surplus), the incentive to free-ride dominates. Indeed, China, whose cooperation is most needed because it has very large, low-cost abatement opportunities, has the strongest incentive to free-ride. The underlying reason for the failure of cooperation is the asymmetry in

[23] Similar results occur in other studies. For example Eyckmans and Tulkens (2003) find that the gain in discounted consumption from full cooperation over the non-cooperative outcome is only 0.5 percent.

net benefits among the six regions. However, all is not lost. The article goes on to demonstrate that with appropriate transfers, a number of stable coalitions can be reached. The one promising the highest welfare increase consists of all regions save Japan and the FSU. Ninety-two percent of the potential welfare gains from full cooperation are captured and 82–83 percent of the environmental benefits are secured. The study does not report the size of the transfers needed to attain this result.

Weikard et al. (2006) divide the world into 12 regions, which gives 4,084 possible coalition structures. The time horizon is 100 years. They are especially interested in how the rules for sharing the coalition surplus impact stability. Stability is modeled using an empirical module that calculates benefits and costs for the twelve regions and applies a game theoretic module to determine stability. Critical empirical assumptions are as follows. China, the United States, and the Former Soviet Union have the lowest marginal abatement costs. Linear benefit functions (i.e., constant marginal benefits) are assumed.[24] This is important because it means non-participants have a dominant strategy and do not adjust their abatement levels to changes in abatement by the coalition. Thus a dominant strategy implies no carbon leakage as the result of an IEA. Three of the twelve regions (United States, Japan, EU) account for 64 percent of global discounted damage costs. This too is important. A common view is that developing countries will suffer the most damages with global warming. This is correct relative to their GDP, but the industrial countries have far more economic assets at risk. The high damage shares and high marginal abatement costs provide them the motivation for cooperation. The base discount rate is set at 2 percent.

The focus in this study is on different rules for sharing the coalition surplus and the impact of these rules on coalition stability. Eight sharing rules are considered: (1) egalitarian (one region one share); (2) shares proportional to regional GDP; (3) shares proportional to regional population; (4) ability to pay – shares proportional to regional per-capita income with lower incomes receiving higher shares; (5) shares proportional to current emission levels, which is basically grandfathering; (6) shares proportional to the inverse of current emission levels,

---

[24] This is modified in their sensitivity analysis.

which is an imperfect measure of historic responsibility[25]; (7) shares of discounted global damages, which compensates the most vulnerable; and (8) shares of aggregate abatement costs, which compensates for greater abatement effort. Schemes based on population, ability to pay, and the inverse of current emission levels reflect a desire for equity in the sharing of the surplus.

The results are as follows. First, the Grand Coalition is not stable under any of the eight surplus-sharing schemes. Second, of the 4,084 potential groupings, 18 stable coalitions consisting of two or more regions were identified. All eight sharing schemes had one or more stable coalitions. China was a member of seventeen out of the eighteen. Low marginal abatement costs make it an attractive partner. The majority of the stable coalitions were small, consisting of only two regions – China and the partner that was favored by the particular savings rule under consideration. For example, under the share-of-damages rule, in which rich countries with large assets at risk claim a large share, China would join with either the United States or the EU, but not both. Under the population rule, China would join with a group of energy-exporting countries, which include populous countries such as Mexico and Indonesia (Russia is in a separate region). Third, for the eighteen stable coalitions, the average ratio of net benefits received by the coalition members versus benefits received by non-members – the external benefits of the coalition abatement – was less than 8 percent. This reflects both the small size of most coalitions and the extreme pull of free-riding. Fourth, the coalition that performed best, as measured by the percent of the Grand Coalition surplus it secured, consists of the United States, Eastern European Countries, Energy-Exporting Countries, and China – a colorful mix indeed. This group captured 36 percent of the Grand Coalition surplus and about 40 percent of the tonnage of emissions reductions that would have been realized in the Grand Coalition. Still, this best-performing coalition manages to let 82 percent of the global benefits it creates escape to free-riders.

---

[25] It is unclear why this metric was chosen. Although there are controversial elements in quantifying historical contributions, considerable work has been accomplished. For discussion see Dellink (2009). One contentious point is how to treat land use changes (deforestation). If included the contribution of developing countries rises substantially. A second contentious point is whether to measure emissions on a production or consumption basis. See Chapter 7.

Finally, with respect to the impact of different sharing rules on the size and effectiveness of coalitions, the study concludes that the obvious equity-based criteria – population, ability to pay, and historic responsibility – performed relatively poorly. Grandfathering, which may be the most inequitable rule, trumps all other schemes in terms of global welfare and emission reductions.[26] This is a troubling result, as developing countries have stressed that post-Kyoto negotiations need to consider the ability to pay and need to redress inequities in current and past emissions.[27] It is also troubling in light of our earlier conclusion that efficiency requires a disproportionate share of permits to be allocated to poor countries.

Bosetti and her colleagues (2009) take a somewhat different approach. They seek out all politically relevant coalitions that are capable of delivering a moderately ambitious emissions target. Politically relevant is taken to be coalitions that include all industrialized countries. The target is an emissions path leading to stabilization of greenhouse gas concentrations at or below 550 ppm CO_2e by 2100.[28] The central estimate is that this will increase global temperature by about 2.5°C. They analyze the internal stability of these politically relevant and potentially effective coalitions, and then look for financial transfer schemes that would make them resistant to free-riding incentives.

The analytical apparatus relies on WITCH, a twelve-region IAM model that incorporates a game theoretic structure in which regions play a non-cooperative Nash game. To incorporate uncertainties, four scenarios are considered with high and low damages and high and low discount rates (pure rate of time discount set at 3 percent and 0.1 percent). *A priori*, taking account of expected abatement costs and damage functions, China, Russia, and Middle Eastern countries appear to have little incentive to join a coalition. Russia can expect

---

[26] This result is consistent with Altamirano-Cabrera and Finus (2006), who examine the effect on coalition formation of rules governing the allocation of permits in a cap and trade system.

[27] Bosello et al. (2003) find that equity based distribution rules can contribute to participation and success of climate change coalitions. But they limit themselves to the distribution of the cost of abatement and do not consider the distribution of the benefits from abatement. Because benefits vary widely they are likely to influence the decision to participate in an IEA.

[28] To accomplish this requires cutting global emissions about 25% below their 2005 levels by 2050 and 50% by 2100.

only modest climate-related damages, and perhaps gains in agriculture. Middle Eastern countries face terms-of-trade losses from cutting back on fossil fuels. And China, because of its assumed low marginal abatement costs, would bear a large share of abatement costs unless compensated with transfers.

The results are discouraging. The authors find that only coalitions including all major emitting countries, including India and China, are technical capable of meeting the targets. Even then, coalition members would have to bring their emission levels down to zero and non-members could not increase their emissions above BAU levels as a result of carbon leakage. The list of potentially efficient coalitions is further winnowed down by accounting for free-rider incentives through the game theory module. The results show that only the Grand Coalition, consisting of all countries, is technically capable of tracing out an emissions path that meets the 550 ppm CO_2e concentration target. But neither the Grand Coalition nor smaller environmentally significant coalitions appear stable in the absence of transfers.

The next step was to see if transfers could stabilize the Grand Coalition and a selection of other environmentally significant coalitions. No set of financial transfers were found that would sufficiently offset free-riding incentives and make either the Grand Coalition or other potentially environmentally effective coalitions internally stable. Although the Grand Coalition could generate a surplus and compensate the net losers in the coalition, the surplus would not be sufficiently large to also overcome free-rider incentives.[29] This reveals the heart of the dysfunctional cooperation market. Although the Grand Coalition would increase world welfare over the non-cooperative solution, it is not self-enforcing.

The analysis does, however, shed light on the importance of distribution rules. In a related work that relies on the same WITCH model, Burniaux and colleagues (2009) study a variety of schemes for distributing emission permits that would meet the 550 pp CO_2e target. As expected, whether a country is a buyer or seller of permits is determined by the distribution rule. That, in turn, determines whether the country enjoys net gains or suffers loses. For instance, by 2050, major

---

[29] China, Middle East and North Africa, and Russia are shown to have the highest incentives to free ride.

Annex 1 countries and regions (United States, Western Europe, Japan, Korea, and Russia) were experiencing annual losses in consumption from four distribution schemes based on population, ability to pay, historic responsibility over the 1900–2004 period, and a scheme that would allocate non-Annex 1 countries sufficient permits to meet their BAU projections. Consumption losses are measured as the gap between consumption under the distribution scheme in question and what consumption would be if permits sufficient to reach the 550 ppm CO_2e were simply auctioned off. The hardest hit was Russia (actually non-EU Eastern Europe) under the historic responsibility and BAU schemes. The same Annex 1 regions, however, made net *gains* from the grandfathering scheme as at least some would be net sellers. Non-Annex 1countries generally gained as compared to an auction of permits, with the largest gains arising from the historical responsibility and BAU schemes. The exception was Africa, which saw its largest losses under the grandfathering scheme and its largest gains under the ability-to-pay scheme. None of this, of course, demonstrates that transfers, in the form of biased permit distribution schemes, can overcome free-rider problems, but it does demonstrate that distribution will be a critical feature of any large-scale permit-trading arrangement.

The Dual Role of Transfers

The conclusions thus far are not especially sunny. There are major asymmetries in cost and benefits among countries, complicating the willingness to cooperate. Full cooperation – the Grand Coalition – appears out of reach even using financial transfers or biased distribution of emission permits in a cap-and-trade system. Coalitions that are stable (self-enforcing) tend be quite small and anemic in securing environmental objectives. The Coase solution of establishing property rights and making a market between those on opposite sides of an externality can falter on the question of rights and on untamed free-riding. Surplus-sharing rules that appeal to equity or historical responsibility may founder and have a perverse effect on participation and efficiency.

We can dig a little deeper into why, despite the ecological surplus, IEA models reach these gloomy results. The distribution of abatement costs is being asked to do two jobs. First, as discussed earlier in this chapter and in Chapter 3, an appropriate distribution supports efficiency by recognizing international differences in marginal utilities of consumption. This implies financing abatement using the Lindahl pricing approach of willingness to pay, which is firmly grounded in equity principles. Second, the distribution of abatement costs via transfers is seen as a tool for discouraging free-riding and for increasing participation in and the effectiveness of an IEA. The first role addresses international disparities in income. The second uses transfers as carrots for signing on.

The conflict between these two roles can be seen by comparing a situation where transfers can be made within a uniform tax or cap-and-trade system, and a situation where transfers are not possible. In the latter situation, optimal taxes are *not* harmonized across countries, but reflect per-capita income differences and hence willingness to pay for preventing climate change. Anthoff (2009) finds that if transfers can be made, and if full participation is *assumed*, not demonstrated, the harmonized tax rate that equates marginal abatement costs is $23 per ton carbon for 2005. But if transfers are not possible, the optimal taxes range from $2 for Sub-Saharan Africa to $12 for China to $137 for the United States and $179 for Japan.[30] Interestingly, emissions reductions below the BAU baseline are about the same with and without transfers. While weighting the damages to poor countries tends to increase stringency of policy, obliging them to pay for their abatement, as is implicit in the no-transfer case, tends to support a less stringent policy. For us, the important conclusion from this exercise is a massive job that transfers would have to perform to equalize taxes and marginal abatement costs across countries at very different income levels.

Jacoby and his colleagues (2008) provide additional insight into this question. They assume a cap-and-trade agreement with full participation. The target is set at a 50 percent reduction in greenhouse gas emissions below 2000 level by 2050. They do not consider the benefits of emissions reduction, so the full incentives for participation are not investigated and a benefit-cost (BC) analysis is not possible. Their purpose is to estimate by region the welfare costs and financial flows (transfers) arising from alternative distributional schemes.

---

[30] The analysis uses FUND, an IAM. The base assumption is a pure time preference rate of 1%, and a coefficient of inequality aversion (η) of 1. The results are in 1995 dollars. The standard utilitarian welfare function is used. See also Chapter 3.

Welfare effects are measured against an adjusted BAU reference scenario. We are particularly interested in the scheme that distributes permits according to ability to pay – that is, the inverse of regional per-capita GDP. This comes closest to our concern that the cost allocation reflects differences in marginal utilities of consumption. We are also interested in the scenario that fully compensates non-Annex 1 countries for their abatement costs, including terms-of-trade losses. This scenario is highly unlikely, but sets an upper bound on the costs that Annex 1 countries would be asked to bear, and an upper bound on international transfers.

The welfare and transfers results are surprisingly large.[31] The ability-to-pay distribution scenario shows average consumption losses for 2020 exceeding 4 percent for the seven Annex 1 countries or regions. This rises to more than 10 percent in 2050. The U.S. consumption losses are somewhat below the averages. The same scenario shows non-Annex 1 countries experience consumption changes ranging from minus 19 percent for the Middle East to plus 56 percent for Indonesia. These numbers change to minus 57 percent and plus 63 percent in 2050. This suggests that a simple distribution scheme based on ability to pay can go beyond adjusting for income differences and can become an income redistribution tool itself.

The size of the transfers under this scenario is equally impressive. By 2020, Annex 1 countries are buying net $884 billion (in 2000 U.S. dollars) in permits from non-Annex 1 countries. By 2050, net annual permit trade between the two groups rises to $2,072 billion. Under the unlikely full-compensation scenario, Annex 1 countries would buy net $434 billion in 2020 and $3,314 billion in 2050. Both scenarios, of course, exaggerate the need for transfers, either to induce participation or to offset differences in marginal utility of consumption, as the benefits of abatement are not included. More precisely, the transfers would offset developing countries' gross, not net, abatement costs. Still, the potential size of transfers raises a question about participation by rich countries. We now see the bare bones of the dilemma. For efficiency *and* equity reasons, the distribution of abatement costs should

---

[31] The authors explain this by noting the stringent targets, the inclusion of non-CO_2 gases arising in developing country agriculture, and the tight time frame they employ. Also it is not clear if the permit allocation is based only on the base year, or if it changes to reflect income convergence.

respect differences in per-capita income. It is unclear, however, if it is in the self-interest of rich countries to participate and fund arrangements that do so.

Stepping Back

This analysis may be unduly negative. The basic premise that all countries, even those that are members of an IEA, act in their narrow self-interest may itself be too narrow. Good deeds may inspire more good deeds. Shaming the uncooperative may work. Abatement may be contagious. Some experimental work suggests that voluntary contributions to global public goods may be greater than standard theory indicates (Burger and Kolstad 2009).

Indeed, the whole framework of these models is a bit odd. They assume we have an interest in the material well-being of strangers who will be living in our country hundreds of years from now, and that this interest is sufficiently compelling that we will reduce consumption today. But the framework also assumes we have *no* interest in the well-being of strange strangers living in other countries either now or in the far future. Inadequate as it may be, the existence of foreign assistance programs tends to refute this perspective.

Moreover the models pay little or no attention to the "no regrets" gains from addressing global warming. These include efficiency improvements from eliminating energy subsidies, collateral benefits for biodiversity conservation from reductions in deforestation, first mover commercial advantages in low-carbon technology, and important health benefits from reduced local pollution from burning fossil fuels. These benefits accrue to precisely those countries active in mitigation. Although perhaps not decisive, they whittle away at free-rider incentives and improve the prospects for cooperation.[32]

The analysis of collateral health benefits is particularly interesting but tricky. Unlike global-warming benefits, which must be heavily discounted over many decades, health benefits start to show up almost immediately. The amount of the benefits depends in part on how and where the abatement takes place and in which sectors, as countries differ in their vulnerability to local air pollution and the sector in

---

[32] China's investment in renewable energy technology and commitment to increase forest cover by 40 million ha come to mind.

which it arises. The monetized results also depend on how the value of a statistical life (VSL) is established. Bollen et al. (2009) choose lower than standard estimates of VSL, arguing that local air pollution disproportionally affects older people whose willing to pay less for a reduction in mortality risk should be less because they have fewer years of life left. (The more senior of us may disagree). VSL is also calibrated to per-capita GDP of the region relative to Europe. Thus a premature death avoided in India is worth only a small fraction of a premature death avoided in Europe. Partly as a result of this, by 2020, the monetized value of premature deaths avoided as a percent of carbon prices is four or six times higher in Europe than in China or India.[33] The gap should close over time as the income gap closes. One irony is that if participation by developing countries is delayed, and if OECD countries persevere in their target, abatement efforts and collateral deaths averted will shift northward, and monetized benefits from collateral health effects actually increase! Overall, the conclusion is, however, that collateral health benefits may strengthen the incentive for developing countries such as China to participate. To put the best face on this, one could argue that the much larger near-term gains by OECD countries in collateral benefits might make them somewhat more generous in providing transfers.

Another limit of these models is that by and large, the "carrots" considered are restricted to financial transfers or disproportionate allocation of emission permits in a cap-and-trade regime. In fact, other payment vehicles are being discussed that may sweeten the pot. It is possible to build a model including international trade in which a participating country captures a fraction of the externality gain through terms-of-trade improvement (Cai et al. 2009), or in which a carbon-motivated regional trade agreement can make a small but positive contribution to reducing emissions (Dong and Whalley, 2009). Indeed, Russia was induced to join the Kyoto Protocol in part by concessions on its membership in the WTO. Participation also may be encouraged through joint R&D arrangements, technology-sharing commitments, and financial and technical assistance for adaptation to climate change. Finally the analysis typically assumes our knowledge of abatement cost

---

[33] The authors note that this is a comparison between average VSL and marginal carbon price.

and benefit functions does not change over time. If decisions could be revised in the future on the basis of new information, the added flexibility might make cooperation appear more attractive today.

On the other hand, the cooperation models may be unduly optimistic. The standard assumption is that a coalition equalizes marginal abatement costs among members completely, with no administrative cost, evasion, discounting of traded permits, or other defects. The models pay scant attention to domestic political economy pressures and to what is likely to be a major source of opposition, loss of competitive position in international trade. In practice, exemptions or special treatment of energy-intensive sectors are being actively discussed in many countries. The competitiveness concern is not limited to the West. China in particular is considering how to cushion the effects of a carbon tax on its exports (Laing et al. 2007). India is concerned that if it were to become a seller of permits internationally, it would appreciate its currency and lose its competitive edge – a new strain of the Dutch disease! (See Chapter 7.) Additionally, modeling has been restricted to a relatively small number of countries or regions, often twelve. Coalition theory suggests that the larger the number of players, the greater the incentive for free-riding. If data at a finer level of disaggregation were available, the models might yield even more discouraging results. Perhaps most important, we can expect that there will be net global costs for many decades before net benefits can be seen.[34] Is it realistic to expect countries to sacrifice consumption for the next fifty years to enjoy uncertain welfare gains in the next century and beyond? And is it reasonable to suppose that during the long period of sacrifice, some countries would be transferring money and resources to other countries to secure their participation? It certainly would be unprecedented.

Sticks

If financial or other inducements are not sufficient to gain acceptable levels of cooperation, are coercive measures more effective? The most obvious and likely sanctions are trade-related, especially import

---

[34] Nordhaus (2010) has estimated that the global costs of the Copenhagen Accord through 2055 are almost 6 times the benefits, although with a long enough time horizon the BC ratio would be favorable.

restrictions to deter free-riding and defections.[35] Trade measures to enforce obligations can also be used. Such sanctions are analytically problematic for four reasons. First, they harm the sanctioning country as well as the target country. When modeling IEAs based on national self-interest, there is a tendency to dismiss them as non-credible threats. Second, precisely because they are coercive, they have the potential to exchange one problem – free-riders – for another problem – forced riders. Coercive trade measures can lead to international extortion, not efficiency. Third, to prevent such behavior, countries are bound by World Trade Organization (WTO) rules as to whether and when they can restrict imports. Coercive sanctions to encourage participation or deter defections from IEAs can run afoul of these rules.[36] (See Chapter 7.) Fourth, the issue of trade sanctions to discourage free-riding and non-compliance has been conflated with the response to carbon leakage and competitive concerns generally. The two issues cannot easily be disentangled. Trade restrictions to deter free-riding, if successful, will also reduce carbon leakage, and restrictions to reduce carbon leakage may affect free-riding. Note, however, that trade restrictions to deter free-riding need not be limited to sectors where carbon intensity and carbon leakage are highest. For maximum coercive effect, they may target other sensitive products and sectors. More importantly, the motivation differs – inducing participation and compliance versus managing competitive loss.[37] Having said that, however, politics sees the two as reinforcing responses.

---

[35] National security concerns aside, for political economy reasons countries are more likely to restrict imports than exports. Scott Barrett has made the important point that enforcing a trade agreement with trade sanctions is easier than enforcing a global environmental agreement because the former involves bilateral interests and the later involves multilateral interests via its global public goods character.

[36] Country specific trade *preferences* are also subject to rules but exceptions to the Most Favored Nation Principle have been liberally granted. Trade preferences to induce cooperation are unlikely to meet much resistance.

[37] The use of trade restrictions to achieve international environmental objectives has been relatively rare but occasionally highly controversial. The Basel Convention on Trade in Hazardous Wastes, and the Convention on Trade in Endangered Species (CITES) use trade measures as the essential enforcement tool, but the measures were not used as an inducement to join. Unilateral use of trade restrictions in the controversial tuna-dolphin and turtle exclusion devise cases were motivated by environmental objectives. In contrast, restrictive trade measures were written into the Montreal Protocol at least in part to encourage widespread participation and discourage free-riding.

Barrett (1997) addressed the issue of trade sanctions in IEAs. Using a self-enforcing IEA model linked to a trade model that assumes a particular oligopolistic market structure, he shows that the threat of trade sanctions is sufficient to deter free-riding and the IEA will achieve full participation, subject to the requirement that it contain a minimum-participation clause. The threat itself is sufficient and trade is not restricted. Barrett takes pains to point out that this arrangement may not pass WTO muster, and in any case the results arise from specific assumptions in the analysis and may not generalize. More broadly, it may not be necessary to be explicit about trade sanctions for them to have an impact. For example, some ambiguity about the U.S. position on carbon-related trade restrictions might induce China to accelerate mitigation plans. The larger issue is whether it is desirable to use the trade system to accomplish international environmental objectives. It is attractive to do so as there are few other ways to manage numerous international environmental externalities. However, the international trade system itself is under stress. To take on environmental objectives might overload it.

The Need for Cooperation

The need for an early and widespread agreement, or at least coordinated action, to reduce greenhouse gas emissions rests in part on technical and cost considerations. If mitigation costs are to be minimized, marginal abatement costs must be equalized across countries, sectors, and gases. Both a uniform greenhouse gas tax or a comprehensive cap-and-trade system are well suited to this task. Incomplete or delayed participation shifts abatement from lower- to higher-cost countries and activities, and will increase the costs of whatever target temperature change is selected. Because the costs will increase, the globally optimal amount of mitigation will be less, temperature increases will be larger, and damages greater.[38]

Fairness and equity norms also make early and widespread agreement desirable. The participation criterion discussed in the previous

---

[38] Limiting restrictions to CO_2 also increases costs substantially. For a comprehensive analysis, see *The Energy Journal* (2006) Special Issue: Multi-Greenhouse Gas Mitigation and Climate Policy.

section – self-interest through maximizing national net benefits – may not be enough if countries feel they are badly treated. Perceptions of extensive free-riding by others can sink a deal. Some sacrifice for the public good is expected from all. Recall that in 1997, the U.S. Congress unanimously passed a resolution stating it would not ratify any agreement unless developing countries committed to limiting and reducing greenhouse gas emissions. This was an emotional response, not based on any cost-benefit calculus.

Finally, an inclusive agreement is needed to weaken the competitiveness/carbon leakage issue. Trade competitiveness is a potent force in the domestic politics of climate negotiations. The long history of opposition in the North to environmental legislation based on the "pollution haven threat," and the equally long concern in the South that sees environmental regulations as covert trade barriers, has set the stage for sharp disagreement over carbon leakage and competitive losses. A commitment to an internationally harmonized carbon price would undercut this particular stumbling block.

Cost Considerations

The models all tell the same story. Incomplete and delayed participation are costly. The cost shows up in carbon prices. Bosetti et al. (2009) estimate that by 2050, a coalition that included all countries (the Grand Coalition) would face carbon prices of about $300 per ton of CO_2. If three relatively small regions did not participate – Southeast Asia (which does not include India), Africa, and non-EU Eastern Europe including Russia, the carbon price rises above $2,000 per ton. Seidman and Lewis (2009) estimate that the cost of a 15 percent reduction in carbon emissions would be about $108 billion annually if undertaken by the world's forty-six richest countries, but only half that – $55 billion – if the abatement effort was efficiently distributed among all countries. His results suggest that the rich countries could fully compensate the poor for their abatement expenses and still pocket $55 billion. (This analysis does not consider damages or free-riding incentives). Nordhaus (2008) constructs a participation function to illustrate the costs of incomplete coverage. If countries accounting for 66 percent of world carbon emission are included, the cost is 2.1 times the full 100 percent participation cost. Sixty-six percent was the original Kyoto coverage that included the United States. With

33 percent of global carbon emissions included, the cost is about 7.4 time's full participation. Thirty-three percent is the coverage in 2010 without the United States and including emissions growth in developing countries.[39]

No one seriously expects all countries to impose a uniform carbon tax or an equivalent global cap immediately and secure the least-cost solution. It is therefore useful to investigate what cost penalties are anticipated should some countries delay participation in an IEA or baulk at immediately going to the common carbon price. The answer depends on the extent of incomplete participation, when it will be remedied, and on the stringency of the concentration target. *Ceteris paribus*, the more stringent the concentration target, the earlier mitigation efforts must be undertaken, and the more costly will be delay in obtaining full participation. A tight target reduces flexibility in spreading emissions reductions over time, making the cost of incomplete participation that much higher. Edmonds et al. (2007) finds that with a relaxed CO_2 concentration target of 650 ppm, a delay in participation by non-Annex 1 countries until 2050 increases carbon prices by about 50 percent over what they would be with immediate participation. But a target set at 450 ppm is simply unattainable at any price if the delay extends to 2050. Even if accession by non-Annex 1 countries is brought forward to 2035, carbon prices within the coalition spike at more than $2,500 per ton. Put another way, the global social cost of delaying accession to 2035 is increased by 23 percent over least cost if the target is 650 ppm, but increased by 265 percent for a 450 ppm target. The combination of a tight target and delayed participation not only causes the price of carbon to spike, but greatly increases real costs, as mitigation is restricted to countries with high and rising marginal abatement costs.

The Edmonds study also underlines conflicting interests. Delay in accession shifts the abatement cost burden away from non-Annex 1 countries. Under immediate participation, they would bear 66 percent of the burden of meeting a 450 ppm target (absent any transfers), but only 35 percent if they delayed until 2035. Still, because of the great inefficiency in delayed participation, their actual realized cost would

---

[39] Not only is delay costly, but some long-term temperature targets become infeasible. See O'Neill et al. (2010).

be higher. The larger question is whether the Annex 1 countries would persist with the 450 ppm CO_2 target in the absence of a firm commitment by developing countries to participate.

Even though it may appear tempting for a developing country to delay participation, there are limits. A strong case can be made that countries that intend to ultimately join an IEA and accept its cost-minimizing carbon price should not delay too long before taking anticipatory action. There are three reasons. First, the capital stock in energy-intensive sectors tends to be long-lived (for example, transportation infrastructure, electric utilities). Rather than lock in high carbon-intensive capital that will eventually have to be scrapped or offset by carbon reductions in other sectors, it would be more cost effective to seek low carbon-intensive technology well before actually committing to a global carbon price. Rail transport, nuclear power, and renewable energy are good examples. (This incentive, however, may be obscured by the Green Paradox and near-term depressed fossil fuel prices, as explained in Chapter 5.) Second, anticipatory policies allow developing countries to profit from international research spillovers in low-carbon technologies. Deploying these technologies takes time. Third, focused R&D conducted by the more advanced developing countries may be cost effective. China is reported to spend 1.5 percent of GDP on total R&D and has targeted 2.5 percent by 2020 (Bosetti 2009), a level comparable to the United States and Japan. Some immediate focus on carbon capture and storage technologies and renewable energy may reduce the subsequent cost of implementing a carbon tax or cap in the more advanced developing countries.

Anticipatory actions may complicate establishing a benchmark BAU trajectory. Consider, for example, a proposal to allocate emission permits to developing countries equal to their BAU needs. The purpose would be to induce their eventual full participation in an IEA. The allocation assures them that their development prospects will not be compromised. It also allows them to become sellers in a global permit market if they exploit their low-cost abatement opportunities. In exchange, the country commits to ultimately aligning its carbon price with the global price. The complication is which BAU projection to use: one that does or does not allow for anticipatory action during the interim period? The difference can be quite large. The Bosetti, Carraro, and Tavoni (2009) analysis concludes that using a myopic, no-anticipation assumption in 2020 OECD countries would spend $94 billion on buying emission permits, with 80 percent going to four countries: Brazil, Russia, China, and India. In contrast, if anticipatory actions are factored into the BAU projection, OECD countries would scale back purchases to $23 billion, with 72 percent going to the same four countries. The reason is that in the latter case, the BAU projections would be reduced by the anticipatory actions the sellers undertake in their own self-interest. The absolute numbers would be higher in 2030 due to larger quantity of permits traded and higher carbon prices, but the difference between permit trade values under the two BAU projections narrows. Note that there is a separate issue as to whether the BAU projections do or do not include the effects of carbon leakage from Annex 1 countries.

A Shrinking Target Space

The issue of participation is not only linked to costs but to what is technically feasible. An incomplete agreement closes off certain concentration and temperature targets. Consider the timing problem. There are good reasons for ramping up climate policy at a moderate, deliberate pace. Better information on the extent of harm will become available. Lower-cost technologies for carbon abatement should also appear. A relatively slow pace will limit premature scrapping of physical capital embodied in transportation, structures, utilities, and so forth. The logic of discounting suggests that, *ceteris paribus*, delaying expensive mitigation is desirable. As previously noted, however, there are also substantial costs involved in delay, especially if the delay is the result of incomplete coverage. Moreover, some greenhouse gas concentration and temperature targets become technically unattainable.[40]

Three variables interact. If there are limits on the rate at which greenhouse gas emissions can be reduced, one can construct a relationship between how long emissions reduction is delayed and the maximum atmospheric concentrations and hence temperature increase

---

[40] Bosetti et al. (2009) estimate that by 2050, all coalitions that meet a 550 ppm CO_2e target must include China and India (unless all other developing countries offset the non-participation of one or the other of these two countries). O'Neill et al. (2010) model the midcentury emissions, concentrations, and technology conditions that need to prevail if we are to achieve widely discussed targets such as holding temperature increases below 2°C.

that can be secured. Fixing a maximum rate of emissions reduction and establishing a target concentration will fix the maximum period that a policy of emissions reductions can be delayed. Conversely, if we continue to fix emission reduction rates and specify the length of delay before emissions stabilize and start to decline, we can calculate the peak greenhouse gas concentration levels. Peak concentration levels can then be converted to (probalistic) estimates of temperature increase. Thus although there are real cost advantages to delay, there are lost opportunities also. Delay will take certain concentration and temperature options off the table. Put somewhat differently, either the time window for robust mitigation policy starts to close or the range of achievable targets shrinks. Major uncertainties complicate the analysis. If we require a 90 percent probability of limiting temperature increase to 2°C, the time window is substantially narrower than for a 50 percent probability.

The maximum rate at which carbon emissions can be reduced is both an economic and a technical question.[41] The higher the rate, the higher the cost, but that relation is very speculative. Remember that carbon emissions from fossil fuels *increased* at more than 3 percent per year from 2000 to 2008 before dipping during the recent recession. To halt the growth and to engineer *declines* in emissions will take a major global effort. Mignone and his colleagues (2008) have taken on this three-variable problem. In the first part of their analysis, they assume that when mitigation starts, emissions will be held constant for a decade and then decline by 1 percent per year thereafter. They then compare starting mitigation immediately with delay scenarios of various lengths (up to fifty years). The results show that with no delay, CO_2 concentrations would peak at 475 ppm in 2150, and with a fifty-year delay, the peak would be 922 ppm in 2240. (The current level is about 390 ppm.) Holding concentrations to 450 ppm no longer appears possible even if emissions were to stabilize today.

If the rate of emissions decline is not fixed for technical reasons, but is considered a policy variable, it is possible to tradeoff the rate of decline with the delay in mitigation while holding a concentration target constant. In essence, this is simply back loading the mitigation

---

[41] The principal study in this area is limited to carbon emissions, not all greenhouse gases.

effort. For example, if the target concentration is set at 550 ppm, a delay of fifteen years is feasible *if* the rate of decline is boosted from 1 percent to 1.2–1.55 percent per year. However, the cost of increasing the rate of decline increases exponentially. To achieve a 450 ppm target after a delay of ten years would require a rate of decline in excess of 3 percent per year, which the authors suggest is not feasible.

James Hansen and his colleagues (2008, p. 17), approaching the timing issue from the perspective of climate science, summarize their analysis: "Continued growth of greenhouse gas emissions, for just another decade, practically eliminates the possibility of near term return of atmospheric composition beneath the tipping level for catastrophic effects."

Conclusions

Economic concepts and analysis have contributed to understanding the climate cooperation conundrum but have not solved it. Difficulty in securing cooperation for a robust climate policy is not due to willful obstructionism. There are more fundamental reasons. At one level, uncertainty and long time horizon interact in the political arena to inhibit a strong response. At a yet deeper level, the nature of the challenge – providing a global public good in a context of asymmetric national costs and benefits, wide disparities in income, highly unequal past and prospective emissions contributions, and free-riding untamed by either a supranational EPA or an adequate stock of carrots and sticks – is more than sufficient to explain low levels of cooperation. Despite the welfare surplus that could be gained from cooperation, which is real, and despite the need to hold down costs and keep reasonable temperature targets feasible, finding a cooperative, comprehensive agreement remains a struggle.

References

Altamirano-Cabrera, J. C. and M. Finus (2006). Permit Trading and Stability of International Climate Agreements. *Journal of Applied Economics* IX (1): 19–47.

Anthoff, D. (2009). Optimal Global Dynamic Carbon Taxation. *ESRI Working Paper* 278.

Barrett, S. (1994). Self-Enforcing International Environmental Agreements. *Oxford Economic Papers* 46: 878–94.

(1997). The Strategy of Trade Sanctions in International Environmental Agreements. *Resource and Energy Economics* 19: 345–61.

(2001). International Cooperation for Sale. *European Economic Review* 45: 1835–50.

(2007). *Why Cooperate? The Incentive to Supply Global Public Goods.* Oxford: Oxford University Press.

(2009). Climate Treaties and the Imperative of Enforcement. In *The Economics and Politics of Climate Change*, D. Helm and C. Hepburn (eds.). Oxford: Oxford University Press.

Bollen, J., B. Guay, S. Jamet, and J. Corfee-Morlot (2009). Co-Benefits of Climate Change Mitigation: Literature Review and New Results. *OECD Economics Department Working Papers No. 693.*

Bosello, F., B. Buchner, and C. Carraro (2003). Equity, Development, and Climate Change Control. *Journal of the European Economic Association* 1: 601–11.

Bosetti, V., M. Travoni, and C. Carraro (2009). Climate Change Mitigation Strategies in Fast Growing Countries: The Benefits of Early Action. *Energy Economics* 31 Supplement 2: S144–S151.

Bosetti, V., M. Travoni, C. Carraro, E. De Cian, R. Duval, and E. Massetti (2009). The Incentives to Participate in and the Stability of International Climate Coalitions. *FEEM Nota di Lavoro* 64.2009.

Buchhotz, W. and W. Peters (2007). Justifying the Lindahl Solution as an Outcome of Fair Cooperation. *Public Choice* 133: 157–69.

Burger, N. and C. Kolstad (2009). Voluntary Public Goods Provision, Coalition Formation, and Uncertainty. *NBER Working Paper W15543.*

Burniaux, J.-M., J. Chateau, R. Dellink, R. Duval, and S. Janet (2009). The Economics of Climate Change Mitigation: How to Build the Necessary Action in a Cost-Effective Manner. *OECD Economics Department Working Paper No. 701.*

Cai, Y., R. Riezman, and J. Whalley (2009). International Trade and the Negotiability of Global Climate Change Agreements. *NBER Working Paper* 14711.

Carraro, C., J. Eyckmans, and M. Finus (2006). Optimal Transfer and Participation Decisions in International Environmental Agreements. *Review of International Organizations* 1: 379–96.

Chander, P. and H. Tulkens (1992). Theoretical Foundations of Negotiations and Cost Sharing in Transfrontier Pollution Problems. *European Economic Review* 36: 388–98.

Chichilnisky, G. and G. Heal (1994). Who Should Abate Carbon Emissions? An International Viewpoint. *Economic Letters* 44: 443–9.

Dellink, R. et al. (2009). Sharing the Burden of Adaptation Financing. *FEEM Nota di Lavoro* 59.2009.

Dong, Y. and J. Whalley (2009). Carbon Motivated Regional Trade Arrangements: Analytics and Simulations. *NBER Working Paper* 14880.

Edmonds, J., L. Clarke, M. Wise, and J. Lurz (2007). Stabilizing CO_2 Concentrations with Incomplete Cooperation. *Pacific Northwest National Laboratory* PNNL 16932.

Eyckmans, J. and M. Finus (2007). Measures to Enhance the Success of Global Climate Treaties. *International Environment Agreements* 7: 73–97.

Eyckmans, J. and H. Tulkens (2003). Simulating Coalitionally Stable Burden Sharing Agreements for the Climate Change Problem. *Resource and Energy Economics* 25: 299–327.

Finus, M. (2001). *Game Theory and International Environmental Cooperation.* Cheltenham: Edward Elgar.

Hansen, J. et al. (2008). Target Atmospheric CO_2: Where Should Humanity Aim? *Open Atmospheric Science Journal* 2: 217–31.

Heal, G. (1998). New Strategies for the Provision of Global Public Goods. *Paine Webber Working Paper Series* PW-98–11.

Hoel, M. (1991). Global Environmental Problems: The Effect of Unilateral Actions Taken by One Country. *Journal of Environmental Economics and Management* 20: 55–70.

Jacoby, H., M. Babiker, S. Paltsev, and J. Rielly (2010). Sharing the Burden of GHG Reductions. In *Post Kyoto International Climate Policy*, J. Aldy and R. Stavins (eds.). New York: Cambridge University Press.

Liang, Q.-M., Y. Fan, and Y.-M. Wei 2007. Carbon Taxation Policy in China. *Journal of Policy Modeling* 29: 311–33.

Mignone, B., R. Socolow, J. Sarniento, and M. Oppenheimer (2008). Atmospheric Stabilization and the Timing of Carbon Mitigation. *Climatic Change* 88: 251–65.

Nordhaus, W. (2007). To Tax or Not to Tax: Alternative Approaches to Slowing Global Warming. *Review of Environmental Economics and Policy* 1 (1): 26–44.

   (2008). *A Question of Balance: Weighting the Options on Global Warming Policy.* Yale University Press. Prepublication version at http://www.nordehaus.econ.yale.edu/Balance_prepub.pdf

   (2010). Economic Aspects of Global Warming in a Post-Copenhagen Environment. *PNAS* 107 (26): 11721–26.

O'Neill, B., K. Riahi, and I. Keppo (2010). Mitigation Implications of Midcentury Targets that Preserve Long-term Climate Options. *PNAS* 107 (3): 1011–16.

Pearson, C. (1975). *International Marine Environment Policy.* Baltimore: The Johns Hopkins University Press.

Seidman, L. and K. Lewis (2009). Compensations and Contributions under an International Carbon Treaty. *Journal of Policy Modeling* 31: 341–50.

Sheeran, K. (2006). Who Should Abate Carbon Emissions? A Note. *Environmental and Resource Economics* 35: 89–98.

Shiell, L. (2003). Equity and Efficiency in International Markets for Pollution Permits. *Journal of Environmental Economics and Management* 46: 38–51.

Weikard, H-P., M. Finus, and J.-C. Altamirano-Cabrera (2006). The Impact of Surplus Sharing on the Stability of International Climate Agreements. *Oxford Economic Papers* 58: 209–32.

9

Beyond Kyoto

Almost two decades ago, the UNFCCC set as its primary objective the stabilization of atmospheric CO_2 at levels that avoid dangerous interference in the climate system. That effort is now tottering. The previous chapter establishes both the need for and the difficulties in reaching a comprehensive and effective international agreement on climate change. The time frame for doings so is narrowing. Emissions reduction obligations under Kyoto expire at the end of 2012 and there is no comprehensive and binding arrangement to take its place. Moreover, the window for attaining moderate concentration and temperature targets is closing.

In the period before the 2009 Copenhagen meeting, many proposals for a post-Kyoto agreement were floated. These ranged from a fully elaborated, gradually implemented cap-and-trade system setting forth emission targets for all countries and all decades (Frankel 2010) to a strategy of supporting "climate accession deals" that would contribute to developing countries' interests *and* emission reduction, but which need industrial countries' financial, technical, or administrative support (Victor 2010). All of these proposals need rethinking in light of outcomes at Copenhagen.

From Kyoto to Bali to Copenhagen

Preparations for a post-Kyoto climate regime have their origins in the accomplishments and weaknesses of the Kyoto Protocol (KP).[1] Three

---

[1] For an excellent account, see Trevor Houser (2010).

great accomplishments were the establishment of specific emissions reduction targets for Annex 1 countries, the inclusion of three flexibility mechanisms to assist in fulfilling reduction obligations, and the inclusion of all greenhouse gases, not just CO_2. The numerical targets were a major step beyond the vague goal for Annex I countries to return emission levels to their 1990 levels that was expressed in the earlier UN Convention on Climate Change. By setting these numerical targets for some countries but not others, the KP helped clarify the UNFCCC statement that climate protection would be on "the basis of equity and in accordance with their common but differentiated responsibilities and respective capabilities" (Article 3, UNFCCC). The specific division of responsibility in the KP was not, however, expected to be permanent. One of the three flexibility mechanisms led to the creation of the European Emissions Trading System (ETS), which provides useful evidence for post-Kyoto planning. Another flexibility provision, the CDM, has been successful in engaging the private sector and developing countries in emission-reduction activities.

As is now evident the KP also has major weaknesses. The first is incomplete participation in reducing emissions and no strategy for altering this. The United States failed to participate due in part to competitiveness concerns and the perception of inequitable sharing of burdens. Also, the need to contemplate reducing emissions in rapidly growing developing countries was misjudged. Second, Kyoto commitments have a short five-year shelf life and are scheduled to expire in 2012. KP was always considered a first step, but there were no long-term plans beyond a call for subsequent negotiations. While understandable, the absence of a durable structure compromises long-term planning and investment decisions with respect to energy sources, technology, transportation infrastructure, and in other critical sectors. Third, the monitoring, compliance, and enforcement provisions have been correctly criticized as absent or inadequate. Fourth, the issues of technology development, diffusion, and adoption, as well as the funding for these activities, were neglected.

By the time representatives met in Bali in 2007 to plan for a post-Kyoto regime, two sets of views had emerged. One set favored an expanded an improved version of Kyoto, a Kyoto II, in which at least key developing countries would be gradually drawn into emission-reduction commitments. The basic approach would center on mitigation actions and would continue to have a top-down character. Negotiations could either set quantitative emissions targets or, following proposals by Frankel and others, could focus on formulae for determining abatement obligations and timing. In an international cap-and-trade regime, incentives for participation could be folded into the permit distribution arrangements. Alternatively, negotiations could center on converging to a common international tax and price for carbon. Flexibility in the distribution of abatement burdens could then be achieved by allowing countries leeway in how rapidly they approach the international price. Broad, phased participation to achieve efficiencies in mitigation and minimize the carbon leakage/competitiveness concerns would be the centerpiece of the new Kyoto.

The second set of views was more critical of Kyoto and skeptical of reform. A diverse group of proposals emerged, which might be loosely termed the bottom-up approach. Suggestions included an emphasis on international sector agreements setting sector emission-intensity targets, an emphasis on reversing deforestation, innovations in supporting technology and its dissemination, and radical reform of the CDM. This second set of views was also compatible with what Stavins (2009) has termed a "portfolio of domestic commitments." Climate action goals would be set domestically but could be coordinated and indeed made more efficient by voluntarily linking climate response systems.

The Bali Action Plan kept all these approaches on the table. Adaptation funding, forestry policy, technology, long term mitigation goals and the participation of the industrializing countries in mitigation efforts were all part of the pre-Copenhagen discussions.

From Copenhagen Forward

What are the implications of the 2009 Copenhagen meeting? From the perspective of the economics of global warming, very little changes. All of the complications arising from the unique characteristics of global warming – the time scale, the uncertainty, and the global nature of the problem – are still with us. The limits of benefit cost (BC) as a guide to policy remain, the central role to be played by pricing greenhouse gases in an efficient response continues to be valid, and the challenge of supplying an international public good in a system of sovereign nation states is, if anything, shown to be even more difficult.

From the perspective of policy, however, the available responses to climate change have shifted and narrowed. A Kyoto II arrangement modeled as a more inclusive version of the Kyoto I approach – the international negotiation of comprehensive, binding targets and time-tables under the UNFCCC – is now unlikely in the extreme. Countries announce their commitments but do not negotiate them. And the commitments come wrapped very differently. Some are contingent on domestic legislation (the United States), some are contingent on commitments by others, including financial resources and technologi-cal assistance. Some are percent reductions from emissions in previous years and some are reductions from BAU levels in future years. Some are intensity targets and some are in absolute terms. Some intensity targets are for carbon (China); some include all greenhouse gases (India). A credible common metric to compare absolute, intensity, and BAU-based targets would be useful to compare effort and perfor-mance. Some commitments will be subject to international verification and others will not. In the absence of financial assistance, the miti-gation actions of non-Annex 1 countries will be subject to their own measurement, reporting, and verification. Verification of reductions is politically sensitive, but is of critical importance given the incen-tives to manipulate targets, inflate mitigation efforts, shirk costs, and capture free-rider benefits. None of the pledges are part of an inter-nationally binding treaty with enforcement mechanisms, nor are they likely to become so. Finally, the Copenhagen Accord is silent on emis-sions trading, and at the moment there is no concrete plan for climate actions beyond 2020.

The new style celebrates diversity in packaging mitigation com-mitments. This may have been necessary to coax any commitments whatsoever. As explained in Chapter 6, intensity targets are regarded as protecting national growth prospects. Targets contingent on the actions of others both protect against bearing disproportionate costs and sweeten the pot for others to join the party. And setting reductions from BAU projections without ceding control over these projections has the patina of cooperation but in fact offers an escape clause.

The diversity mode presents two analytical problems. First, how can the various commitments be made comparable in terms of effort, or sacrifice, or cost? Fairness requires some way to measure these attri-butes. The issue of carbon embedded in international trade, discussed

210 Economics and the Challenge of Global Warming

in Chapter 8, adds further complexity. How will fairness be assessed if countries dodge their commitments through exporting carbon-intensive activities? Second, how can these diverse commitments be consolidated so that their combined effects on greenhouse gas concentrations, temperatures, and costs can be estimated? Measuring the likely emission and temperature effects implied by the Copenhagen Accord is difficult, as nothing is known about the path to 2020 (the date around which the commitments center), nothing is certain about the extent, if any, of international emissions trading, and nothing is yet enforceable. Nevertheless, McKibben, Morris, and Wilcoxen (2010) have made the first attempt. Perhaps their most striking conclusion is the disparity in effort assuming followthrough on Copenhagen commitments and no international permit trade. In 2020, Japan would reduce emissions by 48 percent below BAU, the United States by 33 percent, China by 22 percent. In contrast, some countries and regions would be marginally *above* their BAU projections (India 0.4 percent, Brazil 0.6 percent, OPEC 1.3 percent). The change in 2020 GDP relative to BAU ranges from a loss of 6.3 percent for Australia to a gain of 0.7 percent for India. In a separate analysis, Carraro and Massetti (2010) conclude that China and India's 2020 *intensity* targets may not be "binding" – autonomous energy efficiency improvements triggered by long-term price and technology dynamics could make a specific mitigation policy unnecessary. Zang (2010), however, finds that China's intensity pledge, if honored, will go well beyond BAU projections.

The climate impacts of the Copenhagen Accord are even harder to evaluate as not only is the path to 2020 not specified, but the mitigation effort over the remainder of the century is *terra incognita*. There is a general consensus that if the Copenhagen commitments are generously interpreted, meticulously honored, and followed up by serious post-2020 actions, there is a reasonable probability the 2°C target could be obtained. For example, the research consortium AVOID concludes that post-Copenhagen pledges "are likely to be compatible with a long term target of limiting warming to 2°C PROVIDED the reduction pledges are realized by 2020 AND that further significant reductions continue beyond 2020" (Lowe et al. 2010, p. 8; emphasis in the original. See also Houser 2010).

Whether 2°C is too much or too little is not known. Nordhaus (2010), in a recent analysis using his RICE model, finds the maximum optimal temperature increase to be 3°C, occurring in the first half of the next century. In contrast, Hansen et al. (2007) argue for a maximum 1.7°C increase over pre-industrial time, mainly to prevent irreversible ice sheet melting and species losses. The bottom line appears to be that the diversity of commitments, while welcome for their flexibility, simply adds another uncertainty to the underlying unanswered questions about the appropriate target and whether we are on the right path.

Cancun

The 2010 UN Climate Conference in Cancun Mexico adds very little to this story. Expectations were low and were generally realized. The best that can be said is that the atmosphere was more cordial and less confrontational than in Copenhagen. Specific decisions included: (1) a reaffirmation of the 2°C target with the intention to review and consider a 1.5°C target by 2015; (2) a clarification that mitigation measures undertaken by developing countries, and for which international financing is received, would be subject to international verification, the details of which have yet to be settled; and (3) reaffirmation of the Copenhagen commitment to a $100 billion annual Green Climate Fund by 2020. It did not, however, agree on how the money would be raised or spent. At the same time, the meeting put off setting a target time frame for peaking of greenhouse gas emissions and for setting a global emissions goal for 2050. Perhaps more importantly, participants deferred establishing binding emissions targets for a second commitment period under the Kyoto Protocol, should the KP itself survive after 2012.

In the Aftermath: The CDM and Sector Agreements

The failure to agree on a comprehensive, binding mitigation agreement does not mean all negotiations stop. In fact, a bottom-up approach multiplies negotiation opportunities. Reforming the Clean Development Mechanism is one example.[2] The CDM was designed to accomplish very worthwhile goals – to exploit low-cost abatement opportunities in developing countries, to accelerate the deployment of low-carbon technology worldwide, and to encourage developing

---

[2] Hepburn (2009) gives a good review of financing international carbon mitigation, including reforms of the CDM.

countries to participate in the effort to control global warming. It was also designed to promote sustainable development, but this objective seems to have been left at the wayside.

The CDM has developed a constituency in key countries and is unlikely to be scrapped in the immediate post-Kyoto period. However, it has serious weaknesses that need to be addressed (Bushnell, 2010; Keeler and Thompson, 2010). First, the CDM is not designed to reduce emissions, but rather to achieve Annex 1 reduction targets at low cost.[3] Its ability to do even this may be compromised in a number of ways. The central concept of the CDM is additionality. To earn emission reduction certificates, a project must demonstrate that it reductions are additional to those that otherwise would have taken place. Note that both cap-and-trade and CDM require verification of emissions but only CDM also requires establishing a hypothetical baseline. This opens the door for all sorts of manipulation. Emission levels may be artificially raised to subsequently claim reductions. Certain projects may be proposed, for example coal-fired utilities, that would not have been undertaken in any case, and then credit for switching to gas-fired turbines is requested. Collusion between buyers and sellers of essentially bogus emission reductions credits is certainly possible. And the complexity of documenting additionality has slowed an already overloaded administrative process. Transactions costs have become a major obstacle to efficiency due to the CDM's retail, project-by-project approach.

In principle, the CDM should reduce competitiveness and carbon leakage concerns by contributing to the convergence of carbon prices between Annex 1 and non- Annex 1 countries. By placing an opportunity cost on carbon emissions in developing countries – CDM revenues forgone – carbon prices should rise even without any explicit emission restrictions. The actual effect, however, depends on the BAU baseline in use. If the baseline used to calculate additionality already incorporates leakage resulting from Annex 1 countries' restrictions, the measured reductions from a project are too high. Using inflated CDM credits amounts to additional leakage.

There are, in fact, two fundamental problems with the current CDM. The first is that it is project-based, and to capture many major low-cost

---

[3] Double-counting CDM emissions reductions by both the buyer and seller in monitoring Copenhagen Accord commitments should be avoided.

abatement opportunities will require sector and policy measures. There are numerous obstacles to this radical expansion of CDM, and this suggests that a tax or cap-and-trade system might be more effective. Second, the existing CDM is a comfortable and lucrative source of revenue for several developing countries and provides a disincentive for them to undertake formal emission reduction commitments. This is a form of moral hazard. The disincentive would be weakened if graduation to quantitative emissions targets were made a condition for participating in a revised CDM. In spite of these defects, a revised CDM might make a significant contribution if carbon capture and storage (CCS) technology becomes competitive over the next two decades. Countries such as India, with large unmet energy needs, would need financial assistance in buying the technology. The optimal financing might be through a global cap-and-trade scheme, but if that does not materialize and cost breakthroughs in CCS do, the CDM could be the financing vehicle. In this instance, the problem of additionality disappears along with the verification of sequestered CO_2.

Sector agreements can also play an enhanced role in a bottom-up approach. A relatively few sectors contribute a disproportionate share of carbon emissions. They include the power, metallurgical, iron and steel, paper, and mineral products industries, which together account for about 45 percent of global emissions. About half of these emissions arise in Annex 1 countries. Support for sector agreements is based on the premise that it may be easier to negotiate arrangements among the major producing countries when the discussions are more narrowly confined to a specific industry. If all major sectors could be brought within a series of cap-and-trade or harmonized carbon tax arrangements, emission reductions could be very substantial. Even if a full-blown sector cap-and-trade agreement is not possible, a crediting arrangement for reductions below some negotiated baseline would be helpful. The premise that they are easier to negotiate is untested to date, but the idea of sector arrangements is likely play a larger role post-Copenhagen. Voluntary sector guidelines involving low-carbon technology are also likely but lack the motivating power of a price incentive.

Sector agreements could follow either a cap-and-trade or CDM approach. A sector cap-and-trade could involve two or more countries. The larger the membership, the greater the potential gains. In principle, permit trading between sectors could also be allowed, creating greater efficiency. Distributional issues would be present as they are in all cap-and-trade schemes, but the oligopolistic character of many of these industries and the prevalence of state enterprises makes the political economy dynamics of negotiations unique. If extended to all major producing countries, leakage problems, at least from that sector, would be minimal. There would, however, be incentives to inflate the caps, capture a trade advantage, and compromise the environmental integrity of the scheme. At the same time, the use of sector-level trade measures to encourage participation and to enforce an agreement may be more readily available.

Alternatively and similar to the CDM concept, a baseline sector emission level expressed as emissions per unit sector output could be established. Emission reductions below this level could be credited and sold internationally to any country or region that was under a quantity constraint. This might be more efficient than the current CDM, as baselines would not have to be set project by project. Schmidt and his colleagues (2008) have outlined a "no lose" arrangement in which developing countries agree to sector-specific emission targets and sell reductions below targets in an international market. Targets would be set by independent experts, and industrial countries would be encouraged to provide technology for yet more stringent targets. These ideas are attractive, but it is unclear how governments that receive credits for emissions below BTU would transmit the price signal to their firms and actually capture the reductions. And perverse incentives to inflate the baseline and to delay undertaking binding commitments would still be present. Negotiating sector agreements is likely to involve firms directly. This may be a plus because of common knowledge of the industry and emission reduction technologies. But the negotiations might also promote collusion and cartel-like behavior.

Finally, discussions on reducing emissions from deforestation and forest degradation (REDD+) are well advanced and may receive extra support to help salvage the slim outcome from Copenhagen. Recall that REDD accounts for some 15 percent of all greenhouse gas emissions, and success with REDD would be more than symbolic.

Adaptation Funding

At first glance, the willingness at Copenhagen to discuss large sums of money for mitigation and adaptation assistance is a plus ($100 billion by 2020). There are reasons to be cautious, however. Just as additionality is central to the legitimacy of the CDM, additionality is central to these pledges. Ever since the 1972 Stockholm Conference on the Human Environment, rich countries have pledged additional funds if developing countries would join the fight against environmental degradation. Some modest funds have materialized, especially through the Global Environmental Facility, but whether they are in fact incremental is Talmudic. Moreover, experience with development pledges is not reassuring. Solemn pledges for the Millennium Development Goals are a case in point. In any event, mixing adaptation funding with mitigation negotiations is problematic. The countries most in need of adaptation assistance are likely to be very minor contributors to global warming, and adaptation money is poor bait for attracting significant emission reductions.[4] For their part, developing countries need to remain skeptical of receiving large sums for adaptation. As pointed out in Chapter 5, adaptation funding is likely to be mainly local and national as the benefits are primarily local and national. Now that national commitments by key non-Annex 1 countries have been announced, the incentive for actually providing generous adaptation funds diminishes.

A Limited-Ambition Agenda

Stavins may have been correct when he foresaw a "portfolio of domestic commitments" emerging in the post-Kyoto era. However, that portfolio can be made more effective with international coordination. There are three approaches and each requires that there be some tangible followthrough to the pledges for emission reduction that were submitted following Copenhagen.

Using Price Signals

McKibben et al. (2009) note that comparability of effort in emissions reduction encourages countries to undertake actions, but quantitative targets can be an unreliable measure of effort in a volatile economic environment. Moreover, countries are reluctant to accept binding

---

[4] There can be exceptions. On a per-dollar of GDP basis, Vietnam is a significant emissions source but, because of the vulnerable Mekong and Red River deltas, may require large adaptation expenditures.

quantitative limits under uncertainty about future macroeconomic conditions, the evolution of technology, future fuel prices, and so on. Their suggestion is to supplement quantitative commitments with a price signal. Countries would establish a minimum CO_2e price to establish their level of effort, together with a quantitative limit on emissions for, say, the next decade. They would also establish a maximum CO_2 e price (the price collar). The price limits would rise by a previously fixed annual rate. So long as the carbon price stayed within the collar, the quantitative emissions limit remains a binding obligation. But if for unforeseen circumstances the carbon price rose to its maximum due to a growth surge or for other reasons, the price signal would relieve the country of the onerous emissions limit. The central point of the price signal is to convey to others a serious intent to control emissions, while providing an escape clause should the quantitative emission target prove to be unexpectedly costly.[5]

Linking Cap-and-Trade Arrangements

A bottom-up approach encourages national cap-and-trade arrangements. Linking these schemes should be encouraged even when they fall short of a global agreement. Cap-and-trade systems raise questions a little like preferential trade agreements (PTAs): Are they stepping stones or stumbling blocks to a global system? If they are to be stepping stones, they will have to be linked at some point. The economic gains from linkage are that carbon prices and marginal abatement costs are harmonized over a wider region. The gains are larger the greater the divergence in initial carbon prices. This is consistent with trade theory, where trade equalizes relative prices, and also with coalition theory. As with all voluntary trade, both countries can expect to gain – the seller of permits by obtaining a price in excess of its additional abatement costs, and the buyer by obtaining permits below its own abatement costs. The international distribution of the gain depends on the terms of trade – where the carbon permit price settles.

---

[5] Successful negotiations concerning emissions limits and price collars could eliminate the need for international trading of emissions permits, but would not equalized carbon prices internationally. Moreover, it is unclear that carbon prices can provide an unambiguous measure of effort in controlling emissions. A \$20 per ton CO_2 tax in China may require a far greater abatement cost in China than in the EU.

Linking two domestic cap-and-trade systems does not itself reduce emissions, but the cost savings might encourage tighter caps. Again, following basic trade theory, not all agents within a country will gain from linking. Those who had been net sellers of permits in the higher carbon price country, perhaps because of generous domestic allocation, will lose as the price falls, as will agents in the lower carbon price country who were permit buyers. These distributional effects may be important in political economy analysis, and affect whether regional cap-and-trade schemes turn out to be stepping stones or stumbling blocks. Also, the incentives to move from a two-country to a three-country cap-and-trade system may be blocked by the partner that is a net permit seller if the prospective new member has cheap permits to sell. This is not unlike the objections of countries receiving preferential tariff treatment to multilateral trade liberalization, which would threaten their margin of preferences.

There is also a possible parallel with the trade theory concept of trade creation and trade diversion. Consider a national level cap and trade system in country A, which allows a limited amount of permits to be credited under the Clean Development Mechanism. Because of the restriction, these permits sell above the marginal abatement cost in developing countries. Now assume A joins in unrestricted linking with country B that has low-cost permits to sell. A reduces its own abatement and purchases permits from B. This is equivalent to trade creation and shifts abatement to a lower-cost supplier. But A may also shift some of its purchases away from CDM to its new partner. This may be equivalent to trade diversion. If B's abatement costs, while lower than A, are higher that the marginal abatement cost of developing countries, linking shifts abatement to a higher cost location, with a corresponding efficiency and welfare loss. Although this example may sound contrived, the diversity of schemes, some with intensity targets, some with complicated cost containment provisions, and some with offset limitations, suggests that inefficient linking, or linking that increased overall emissions, is not out of the question. Moreover, by leaving the cap to be determined at the national level, progressive linking of systems sets up an incentive to set the cap too high and become a permit seller. The seller is then in collusion with permit buyers who are looking for low prices. A race to the bottom and creation of "permit havens" could follow. These difficulties are not insurmountable but do underline the need to look before you link.

A Spontaneous Emissions Reduction Credit Market?

Is it possible that the elaborate and frustrating process of negotiating a successor agreement to the Kyoto Protocol is largely unnecessary?[6] If we take the current commitments of the industrial countries at face value, and if they were to honor those commitments through their own emissions reductions, the costs would be high and unnecessarily so. They have something to buy – low-cost abatement. Many developing countries with low marginal abatement costs have something to sell – low-cost abatement. The item for market is certified emission reduction credits. To make this market work requires three things. First, the industrial countries have to be firm in their commitment to major reductions over the next one or two decades. Second, developing countries have to be convinced that the price they receive for holding emissions below BAU levels is greater than their cost of doing so. In making this calculation, they need to consider the climate damages they themselves avoid. Third, there needs to be some mutually accepted mechanism for calculating BAU emissions and verifying reductions below this level. If moderately successful, in such a market, greenhouse gas prices will tend to equalize internationally and the carbon leakage issue will diminish. If highly successful, industrial countries may acknowledge lower costs in meeting their obligations and decide they can afford tighter emission standards.

The market does not need to be global to start, but enough developing countries must participate to establish a credible supply of credits for sale. Their incentive to join is the revenue they receive. Adaptation funds, technology transfer commitments, and trade access guarantees may be sweeteners, but the core is to establish an opportunity cost for not participating. Some of the models in the previous chapter suggest that free-rider considerations would limit participation. This may be the case but it would be worthwhile putting that to the test. A wavering of commitments by the industrial countries when the costs pile up seems more likely.[7]

---

[6] Jaffee and Stavins (2010) sketch out bottom-up possibilities.

[7] Voluntary pledges also muddy the offset market. It is difficult to see how a truly voluntary pledge for emission reduction can be converted to a saleable asset.

The mobilization of substantial financial resources through a "Green Climate Fund" as envisioned by the Copenhagen Accord has an ambiguous impact on the prospects for an informal reductions credit market and for real emissions reductions. On the one hand, it could jump-start the market with a large infusion of funds. Carraro and Massetti (2010) demonstrate that a reasonable fraction of the Fund (if it materializes) could finance substantial emissions reductions in developing countries, making the 2°C target more realistic. On the other hand, if the funds simply compensated developing countries for honoring their Copenhagen pledges, the money would not be buying any *additional* emissions reductions. In that sense they would replicate CDM payments. In addition, the prospect of selling additional emissions reductions may discourage some countries from making further voluntary reductions, especially in the post-2020 period. The ambiguity about the true emissions reduction impact surrounding the initial pledges, as discussed earlier, is compounded by ambiguity about what such mitigation payments are expected to accomplish.

Technology Policy

The implications of Copenhagen for technology policy are also mixed. On the one hand, it failed to map out an international agreement that would make rising carbon prices credible over the next decades. Energy conservation, carbon capture and storage, nuclear, and renewable technologies did not receive the long-term price assurances that large investments may need. There are, however, grounds for a more optimistic assessment. All three of the approaches discussed previously – an agreement on using price signals, linked cap-and-trade, and a spontaneous emissions reduction market – have the potential for providing a price incentive *if* there is a followthrough on pledges, and *if* there is early and serious discussion of post-2020 actions. The deployment of those technologies could be financed in part through the Copenhagen Green Climate Fund, a reformed CDM, and an international carbon market.

International technology cooperation would of course be desirable. But enough commercial advantage may accrue to the innovation leaders that the equity and free-rider problems that plague cooperation on abatement burden sharing can be partly sidestepped. Vigorous national-level technology policies, backed by carbon pricing and funds for deployment in poor countries, should be an important part of the post-Kyoto regime.

References

Burniaux, J.-M., J. Château, R. Duval, and S. Janet (2008). The Economics of Climate Change Mitigation: How to Build the Necessary Action in a Cost-Effective Manner. *OECD Economics Department Working Paper 658.*

Bushnell, J. (2010). The Economics of Carbon Offsets. *NBER WP 16305.*

Carraro, C. and E. Massetti (2010). Beyond Copenhagen: A Realistic Climate Policy in a Fragmented World. *FEEM Nota di Lavoro* 136.2010.

Frankel, J. (2010). A Proposal for Specific Formulas and Emission Targets for All Countries and All Decades. In *Post-Kyoto International Climate Policy*, J. Aldy and R. Stavins (eds.). New York: Cambridge University Press.

Hansen, J., M. Sato, R. Ruedy et al. (2007). Human-made Interferences with Climate: A GISS ModelE Study. *Atmospheric Chemistry and Physics Journal* 7: 2287–312.

Hepburn, C. (2007). A Review of the Kyoto Mechanisms. *Annual Review of Environment and Resources* 32 (1): 375–93.

(2009). International Carbon Finance and the Clean Development Mechanism. In *The Economics and Politics of Climate Change*, D. Helm and C. Hepburn (eds.). Oxford: Oxford University Press.

Houser, T. (2010). Copenhagen, the Accord, and the Way Forward. *The Peterson Institute for International Economics*. Policy Brief PB 10-5.

Jaffee, J. and R. Stavins (2010). Linkage of Tradeable Permit Systems in International Climate Policy Architecture. In *Post-Kyoto International Climate Policy*, J. Aldy and R. Stavins (eds.). New York: Cambridge University Press.

Keeler, A. and A. Thompson (2010). Resource Transfers to Developing Countries Improving and Expanding Greenhouse Gas Offsets. In *Post-Kyoto International Climate Policy*, J. Aldy and R. Stavins (eds.). New York: Cambridge University Press.

Lowe, J. A. et al. (2010). Are the Emissions Pledges in the Copenhagen Accord Compatible with a Global Aspiration to Avoid More Than 2°C of Global Warming? *AVOID – Avoiding Dangerous Climate Change, Technical Note.*

McKibben, W., A. Morris, and P. Wilcoxen (2009). Achieving Comparable Effort Through Carbon Price Agreements. *Viewpoints. The Harvard Project on International Climate Agreements.*

(2010). Comparing Climate Commitments: A Model-Based Analysis of the Copenhagen Accord. *The Harvard Project on International Climate Agreements* Discussion Paper 10–35.

Nordhaus, W. (2010). Economic Aspects of Global Warming in a Post-Copenhagen Environment. *PNAS* 107 (26): 11721–26.

Schmidt, J., N. Helme, J. Lee, and M. Houdashelt (2008). Sector-based Approach to the Post 2012 Climate Change Policy Architecture. *Earthscan: Climate Policy* 8: 494–515.

Stavins, R. (2009). A Portfolio of Domestic Commitments: Implementing Common but Differentiated Responsibilities. *Viewpoints. The Harvard Project on International Climate Agreements (October 19)*.

Victor, D. (2010). Climate Accession Deals for Taming the Growth of Greenhouse Gases in Developing Countries. In *Post-Kyoto International Climate Policy*, J. Aldy and R. Stavins (eds.). New York: Cambridge University Press.

Zang, Z. (2010). Assessing China's Carbon Intensity Pledge for 2020: Stringency and Credibility Issues and their Implications. *FEEM Nota di Lavoro 158.2010*.

10

A Summing-Up

Conclusions

There is no "atmospheric economics" distinct from "terrestrial economics." Analyzing global warming requires the standard economic tools. But because of the unique characteristics of climate change – the time frame, the uncertainty, and the global aspects – some tools have been sharpened or redesigned to meet new challenges. The clearest examples are in discounting, policy under profound uncertainty, integrated modeling of economic and environmental systems, environmental policies using market incentives, policies in second-best contexts, and the economics of global public goods. In addition, value judgments and thus ethical issues permeate global warming economics to an unusual extent and are reflected in the literature.

How successful has economics been in answering the three questions that form the structure of the book? The answers are mixed. Benefit cost (BC) is the main approach to determining how warm is too warm. Its origins in building dams and bridges with public funds are far removed from climate change, and its weaknesses in this latest assignment are easy to document. The main ones are the uneasy relations between efficiency and equity, discounting over many generations, which is a novel task for BC, and the limited ability to accommodate profound uncertainty. These weaknesses are compounded by long-standing difficulties in monetizing environmental effects and in using social weighting. The last is of special importance in light of current and prospective inequities in the international distribution of income and the disproportionate damages to be borne by poor countries.

Laying out the weaknesses of benefit cost does not, however, answer the question. Unfortunately there is no other approach that is clearly superior. The precautionary approach is not incompatible with BC analysis. Risk aversion can be accommodated in damage assessment and valuation (crudely, by setting an arbitrarily low discount rate on risk prevention expenditures, preferably using expected utility theory). The precautionary approach also properly calls attention to irreversibility, another hallmark of global warming. Irreversibility can be addressed in principle through the concept of option values, also a standard feature of modern cost benefit. The precautionary approach is designed with catastrophe in mind, an issue with which BC is poorly equipped to deal. But in this critical area, the precautionary approach itself cannot tell us how much to spend on mitigation and how fast. Finally, the closer inspection of the 2°C target approach simply brings us back to the need to consider both costs and benefits.

Despite its multiple frailties, BC is the best economics can do in determining how warm is too warm. Therefore, it is all the more important that the results be presented to policy makers with all the assumptions explicit and with clearly presented sensitivity analysis. It is especially important that the distributional and equity consequences be part of that presentation. These, of course, are the great strengths of good BC analysis – forcing one to think through alternatives and spelling out assumptions and their implications. The downside is that by changing a few key assumptions, very different BC ratios result and very different views on urgency emerge. Because the end product of BC can vary widely, it is relatively easy to stake out quite extreme positions and find support in the numbers. The objectivity of economics as a guide to policy can be compromised. More importantly, there is a legitimate concern that the public attaches undue credence to either inflated cost numbers or inflated benefit (damage) numbers, and major policy errors are made.

The second structural question, concerning tools and strategy, is less controversial. There is almost universal agreement among economists that putting a price on greenhouse gas emissions should be the centerpiece of policy. Ideally the price would be equal across countries, sectors, and gases. Efficiency is the justification. Pricing pollution is not a new tool, but the importance of fossil-fuel-based energy in the economy imparts novelty. Within this broad consensus, there are numerous

differences, some with substantial efficiency and distributional consequences. The merits of cap-and-trade schemes versus tax schemes (and
hybrid combinations) is one. The need to supplement a pricing policy
with explicit technology policies, often subsidies, is another. Policies
designed to moderate carbon leakage and competitive effects are a
third. There is no need for a full listing of what has been presented earlier. The point is that economics has made solid advances in these and
related areas, especially policy interactions with multiple distortions
(e.g., the double dividend and tax interaction effects) and policy under
uncertainty. Economics can also take a large measure of credit for the
flexibility mechanisms in the Kyoto Protocol and the introduction of
the European carbon trading scheme.

Economics has not demonstrated that a mitigation strategy always
dominates an accelerated development strategy. The latter enhances
adaptive capacity and moderates damages when they materialize. It
can also be argued that funds devoted to rapid development will generally earn higher returns than funds devoted to mitigating global
warming, leaving poor countries with greater future wealth. Still, rapid
economic development under the existing inadequate climate regime
accelerates global warming, and there is no guarantee that funds not
spent on global warming abatement will be made available for accelerated development. In any event, this may be a false choice. Even
models that use a high discount rate reflecting returns from alternative investments conclude that there is a case for positive and rising
carbon prices.

Although at a high level of abstraction, mitigation and adaptation
can be considered substitute responses to global warming, they are
fundamentally different. Mitigation attempts to maintain a global
public good by reining in a transnational externality – greenhouse gas
emissions. Adaptation is taking defensive measure against temperature increases. If mitigation fails, adaptation is the default strategy. The
provision of public goods may be part of adaptation (e.g., sea walls),
but unlike mitigation, they are local or national, not international.
Adaptation decisions require good BC analysis but pose no novel
challenges to economic analysis. Adaptation and mitigation are both
required and the economic task is to coordinate an efficient balance.

Technology will be key to achieving mitigation at reasonable
cost. Economics has contributed by examining interactions between

greenhouse gas pricing policies and technology, and by evaluation of alternative technology development and deployment policies. The economics of breakthrough technologies such as carbon capture and storage will be important. The economics of radical technological solutions – geo-engineering – may also flourish. Having said this, our understanding of guiding and promoting technology remains imperfect and may stumble on the challenges of global warming.

The third structural question is the contribution economics can make to cobbling together effective global climate agreements or actions when the players are sovereign states. The jury is still out as to whether there will be such agreements, and whether economics makes much of a contribution. There is no question that economic concepts help clarify the challenge. The notions of global public goods and bads, differences among countries in the marginal utility of income, self-enforcing agreements, and free-riding and strategic behavior certainly all contribute to our understanding. The main tools economics employs are game and coalition theory married to Integrated Assessment Models. The results have not been encouraging. Still, the modeling is primitive and speculative and may be unduly pessimistic. Domestic political economy pressures from interest groups are generally missing (but could work in either direction). The premise that countries are immune to shaming, herding, and the plight of others outside national borders can be questioned. Coercion has yet to be employed (and perhaps should not be). In short, economics can greatly help in understanding the difficulties in capturing the promise that cooperation holds out, but it has no silver bullet to accomplish this. Even without a comprehensive agreement, initiatives undertaken under the more limited agenda described in the preceding chapter will benefit from careful economic analysis. These include codifying various emissions reduction pledges, linking cap-and-trade systems subglobally, price "collars" to reduce cost uncertainty, technology development and sharing agreements, reforms of the CDM, and sector initiatives.

Prospects for "Atmospheric Economics"

Climate change sits firmly within the field of environmental and natural resources economics and is unlikely to migrate or spawn a new subfield. But there is plenty of work to be done. The answers to our first and third questions remain unsatisfactory. We can speculate that advances will be made across many issues: theory and practice of declining discount rates; replacing the underspecified discounted utility model and freeing up η from her multiple roles; linking climate change and sustainability more clearly; thinking deeply about deep uncertainty and catastrophe; improving monetary damage estimate, especially for ecological assets and for social structures; planning adaptation strategies; and analyzing the interactions of international trade in emission permits with trade in goods and with international capital flows. In particular, we need to think more and think better on how to cooperate and unlock the "ecological surplus" hidden within a serious global climate agreement.

Prospects for Climate Policy

It is not the purpose of this book to analyze current or prospective negotiations. Still, the study would be incomplete without expressing an opinion. First, it seems obvious that there will be precious little progress internationally unless and until the United States makes a firm, credible commitment through legislation to substantially cut greenhouse gas emissions. Second, there are serious questions whether the UNCCC is the appropriate venue for mitigation negotiations. A smaller group may be more productive. Third, a top-down, expanded, and enforceable Kyoto II seems beyond reach at this point. In that event, building through a bottom-up process is essential. Fourth, the need for a major technology push in renewable energy and in carbon capture and storage, and in the global deployment of this technology, should be part of this bottom-up process. Carbon pricing alone may be inadequate. Whether geo-engineering is part of that effort is moot. Fifth, realistically, if regretfully, planning adaptation strategy takes on greater importance. Finally and above all, it would be productive to recast the climate debate from a zero-sum game to a positive-sum opportunity. This suggests thinking less in national interest terms and more in terms of fairness to the many generations to come, regardless of their nationality.

Index

adaptation: anticipatory vs. responsive,
101; capacity for, 97; cost estimates
(agriculture, health, infrastructure),
87–90; and development, 97–9;
empirical studies, 104–6; funding, 214;
vs. mitigation as imperfect substitute,
100–4; vs. mitigation for funding, 101
additionality, 212
Agrawala, S., 105
agriculture: crop model approach,
82–3; loss estimates, 13, 83; Ricardian
approach, 81–2
Aldy, J., 25, 152
Altamirano-Cabrera, J. C., 187
Anthoff, D., 59, 64, 66, 68, 69, 85, 105, 190
Archer, D., 10, 45
Arrow, K., 30
Arrow-Lind Theorem, 29
Atkinson, G., 58
atmospheric economics, 222, 225
Azar, C., 66

Babiker, M., 15, 152
Baker, E., 115
Bali, 16
Bali Action Plan, 208
Barker, T., 151
Barrett, S., 118, 176, 179, 180, 181, 182,
183, 196
Baysian learning, 33
Belli, P., 30
Beltratti, A., 36
Birdsall, N., 48
Boardman, A., 47
Bollen, J., 193

border tax adjustments: background,
153–4; legal aspects, 157–8;
quantification, 156–7
Bosello, F., 80, 106, 187
Bosetti, V., 11, 116, 183, 187, 197, 199, 200
Bovenberg, L., 137
Brekka, K. A., 65
Brenton, P., 159, 160
Brovkin, V., 10, 45
Buchholz, W., 175
Burger, N., 192
Burniaux J., 139, 140, 166, 167, 188
Bushnell, J., 212

Cai, Y., 193
Cancun, 17, 211
carbon capture and storage, 109, 111, 112,
117, 120, 140, 213, 219
carbon embodied in trade: estimates,
161–2; policy implications, 162–4;
UNCCC accounting method, 160–1
carbon equivalent of CO_2, 10
carbon equivalent warming potential
(CO_2e), 11
carbon fertilization, 81, 83
carbon labeling, 147, 158–60
carbon leakage: and CDM, 212; channels,
150–1; estimates, 151–3
carbon price path, 107–8
carbon tax: 107–10. *See also* policy tools
Carraro, C., 183, 199, 210, 219
catastrophe, 31–4
Chakravorty, U., 115
Chander, P., 178
Charnovitz, S., 157

Chichilnisky, G., 36, 146, 175
Clean Development Mechanism (CDM):
    15, 211–13; goals, 211–12; limitations,
    212–13; as subsidy, 139
climate sensitivity, 12, 27, 74, 76
Cline, W., 48, 81, 85
Coase Theorem, 130, 171
collateral benefits: health, 192–3;
    other, 192
competitiveness. *See* carbon leakage
Copeland, B., 144, 146, 149
Copenhagen Accord, 16, 208–11
Copenhagen pledges, 17, 210
Corfee-Morlet, J., 77
Cowell, F. A., 57
Crutzen, P. J., 118

damages by type/sector, 13–14
damage estimates: agriculture, 83; sea
    level rise, 84–6
damage valuation issues, 79–81
Dasgupta, P., 50, 51, 57
Dasgupta, S., 84
de Bruin, K., 101, 105
Dean, J., 145
Dellink, R., 105, 186
Dietz, S., 33, 51, 56
diminishing marginal utility, 52–5, 64
discount rate parameter rho (ρ), 49–52
discount rate parameter eta (η):
    estimating, 57–9; meaning, 53;
    three roles, 43–4, 53, 55–7
discount rates: descriptive vs. prescriptive,
    45–8; hyperbolic (declining), 61–3;
    importance, 45; negative, 44; Ramsey
    equation, 48; and sustainability, 34–6;
    and two sector models, 35–6, 61; and
    uncertainty, 62
discounted utilitarianism, 43, 49
Dismal Theorem, 33
Dong, Y., 193
double dividend, 132, 135, 139
Dowlatbadi, H., 104
Dutch disease, 164–7, 194

eco-labeling. *See* carbon labeling
ecological surplus, 176, 178
economic development, 97–9
Edmonds, J., 198
Eichner, T., 108
Ellerman, D., 127

emission targets. *See* targets for mitigation
Environmental Kuznets Curve (EKC), 145
equity weighting. *See* social (equity)
    weighting
European Trading System (ETS), 15, 207
Evans, D. J., 58, 60
expected utility theory, 29
Eyckmans, J., 183, 184

Fankhauser, S., 57, 64, 67
Farrow, S., 22
Farzin, Y.H., 108
fat tails, 33
Fell, H., 134
Feng, Y., 38
Finus, M., 183, 187
Fischer, C., 137, 140
Fisher, A., 30, 35
forestry: and Green Paradox, 111; REDD,
    91–5, 214
Fox, A., 137, 140
Frankel, J., 206
Füssel, H.M., 10

game theory, 187–8
Gardiner, K., 57
Garnaut, R., 12, 84
geo-engineering, 97, 117–19, 121, 225
Gerlagh, R., 115
Global Environmental Facility (GEF), 177
Glombek, R., 150
Goulder, L., 115, 130, 137, 153
Green Climate Fund, 101, 211
green golden rule, 36
green paradox, 106–11, 120
greenhouse gasses: concentrations, 11;
    lifetimes, 10; listed, 10
Groom, B., 63
Grossman, G., 144, 145
Guo, J., 62, 63

Ha-Duong, M., 60
Hanemann, W., 82
Hansen, J., 202, 211
Harrod, R., 50
Harstad, B., 108
Hart, R., 115
Heal, G., 35, 36, 50, 174, 175
hedonic wage models, 25
Helm, C., 167
Helm, D., 161, 163

Hepburn, C., 16, 63, 64, 66, 68, 211
Ho, M., 152
Hoel, M., 35, 61, 109, 150, 177
Hope, C., 59, 76
Hotellings Rule, 111, 112
Houser, T., 206, 215
Howarth, R., 36
Hufbauer, G., 157
Hummels, D., 148

inequality aversion (egalitarian
preferences), 44, 55–7, 60, 65, 183
Inglesia, A., 82
Integrated Assessment Models (IAMs),
21, 26, 73–6, 183
inter-generational transfers, 21–3
Intergovernmental Panel on Climate
Change (IPCC), 9, 13, 14
International Environmental
Agreements (IEAs): defections, 179,
181; grand coalition, 179, 184, 186, 188;
incomplete participation, 196–200;
and leakage, 181; self enforcing, 179,
180; simulations, 183–9; stability of
179; surplus sharing rules, 185–7; trade
sanctions for IEAs, 196; and transfers,
184–5, 187–9
irreversibility, 26, 27, 30–1
iso-elastic utility function, 54, 56, 60, 66

Jacoby, H., 190
Jaffe, E., 111, 167, 218
Jamet, S., 77
Johansson-Stenman, O., 65

Kaldor Hicks hypothetical compensation
test, 21–3, 39, 47
Karp, L., 127
Kaya identity, 26
Keeler, A., 212
Kim, J., 157
Kinderman, G., 94
Kolstad, C., 192
Koopmans, T., 51
Krueger, A., 144, 145
Krutilla, J., 35
Kuznets, S., 145
Kverndokk, S., 64
Kyoto Protocol: accomplishments, 206–7;
flexibility mechanisms, 15, 20; targets,
15; weaknesses, 207

Laing, Q-M., 194
Lenton, T., 12
Lerner Symmetry Theorem, 156
Lewis, K., 197
Lindahl prices, 69, 174–5, 190
linking cap and trade systems, 216–18
Lowe, J. A., 210

Manne, A., 48
Massetti, E., 210, 219
Mathai, K., 115
Mattoo, A., 155, 156, 166
McKibben, W., 210, 215
Meadows, D., 34
Meinshausen, M., 39
Mendelshon, R., 13, 48, 65
Milliman, S., 114
Mingone, B., 201
Mishan, E., 23
Monte Carlo analysis, 28–9
Montreal Protocol, 180
Morris, A., 210

Nakano, S., 162, 164
Nelson, G., 87
Neumayer, E., 35
Newell, R., 62, 111
Nicholls, R., 85
Nordhaus, W., 15, 45, 48, 50, 57, 98, 133,
173, 194, 197, 210
Norgaard, R., 36
North American Free Trade Agreement
(NAFTA), 144

Ocean Dumping Convention, 178
OECD Guiding Principles, 143
O'Neill, B., 198, 200
option values, 30, 38

Pan, J., 161
Pareto optimum, 21, 23
Parry, I., 130, 136
Parry, M., 90
Pearce, D., 46, 57, 77
Pearson, C., 143, 153, 178
Perrings, C., 103
Perrson. T., 12, 26
Perrson, U. M., 35
Peters, W., 175
Pethig, R., 108
Petschel-Held, G., 37

Pezzy, J., 136
Pfeffer, W., 13
Pigou, A., 50
Pigouvian tax, 59, 78, 115
Pindyck, R., 30, 33, 34
Pizer, W., 62, 126, 133, 152
policy tools: advantages of market based,
    131–2; and carbon leakage, 133; and
    carbon tax rebates, 136; and green
    paradox, 109; international aspects,
    137–8; subsidies, 138–40; tax vs. cap and
    trade, 109, 133–4; uncertainty, 134
Polluter Pays Principle (PPP), 143, 175
pollution havens, 145–6, 197
Popp, D., 111, 113, 116
Portney, P., 48
precautionary principle/approach, 30, 33,
    37–8, 49, 223
Prince, R., 114
public goods/bads: attributes, 172; global,
    172; and market failure, 172–3; optimal
    supply, 173–4

radiative forcing, 9
Rahmsdtorf, S., 83
Ralb, A., 28
Ramanathan, V., 38
Ramsey, F., 48, 50
Ramsey Equation. *See* discount rates
REDD. *See* forestry
Rehdanz, K., 167
Ricardian approach. *See* agriculture
Rinaud, J., 152
risk and uncertainty: 20–1, 25–31, 62;
    clustered, 26–7; re intensity targets,
    126–7; techniques to accommodate,
    28–30
risk aversion, 29–30, 38, 53, 56, 59, 60, 223
Rose, A., 64
Rosenzweig, C., 82

Saelen, H., 56
safe minimum standards, 37
Schellnhuber, T., 39
Schmidt, J., 124, 214
Schuster, U., 10
sea level rise, 13
second best, 137, 140, 151
Seidmen, I., 197
sensitivity analysis, 2
Sezar, H., 58, 60
Sheeran, K., 175

side payments. *See* transfers
Sinclair, P., 108
Sinn, H-W., 109
Smulders, S., 108
social cost of carbon, 59–60, 63, 75
social (equity) weighting: 25, 75, 223;
    concepts, 63–6; and financing, 67–9; in
    practice, 66
social utility function, 43, 53
Solow, R., 50
Squire, L., 66
Stavins, R., 167, 208, 215, 218
Steers, A., 48
Stehr, N., 99
Stern, N., 48, 50, 51, 57, 77
Sterner, T., 35, 61
Stockholm Conference, 143
Storch, H., 99
Strand, J., 109
supplementarity, 15
sustainability, 34–6

Tahvonen, O., 108
targets for mitigation: absolute vs.
    intensity, 124–9; post-Copenhagen, 17,
    209; shrinking window, 200–2
Tavoni, M., 199
Taylor, M. S., 144, 146, 149
technology: empirical studies, 115–17;
    endogenous vs. exogenous, 112–13;
    implications of Copenhagen; and
    mitigation, 114–15
Teller, E., 118
Thomas, C., 14
Timilsina, G., 125
Tinbergen, J., 112
Tol, R. S. J., 13, 32, 38, 59, 64, 66, 68, 78, 81,
    85, 99, 104, 167
tolerable windows approach, 37
trade and warming: analytical
    approaches, 144–7; green trade, 147;
    permit trade restrictions, 167–8;
    transportation, 147–9. *See also* border
    tax adjustments; carbon leakage
transfers (side payments), 175, 178, 179,
    182–3, 189–92
Treich, T., 60
Tulkens, H., 101, 178, 183, 184

Ulph, A., 108
Ulph, D., 57, 108
uncertainty. *See* risk and uncertainty

United Nations Framework Convention
on Climate Change (UNFCCC), 14, 20,
37, 176, 206, 207
utility functions. *See* social utility function

value of statistical life (VSL), 25, 193
van der Werf, E., 108
van Steenberghe, H., 101
Victor, D., 118, 119, 206
Viscusi, W. K., 25
von Below, D., 12, 26

Watson. A. J., 10
Weart, S., 14
Weikard, H-P., 185

Weitzman, M., 33, 37, 52, 56, 57,
62, 78, 134
Weyant, J., 48
Whalley, J., 193
Wilcoxen, P., 210
willingness to pay/accept (WTP/A), 23–5,
77, 174
Wing, S., 127
World Trade Organization (WTO),
157–8, 178

Yohe, G., 32, 59

Zang, Z., 210
Zhao, J., 127